**MATHS IN ACTION**

# ADVANCED HIGHER

# Statistics 1

Ralph Riddiough
John McColl

This edition first published in 2000 by:
Thomas Nelson and Sons Ltd

Reprinted in 2001 by:
Nelson Thornes Ltd
Delta Place
27 Bath Road
CHELTENHAM
GL53 7TH
United Kingdom

01 02 03 04 05 / 10 9 8 7 6 5 4 3 2

A catalogue record for this book is available from the British Library

ISBN 0-17-431544-9

Typeset by Upstream, London

Printed and bound in Croatia by Zrinski d.d. Čakovec

# Contents


| | | |
|---|---|---|
| **Preface** | | v |
| **Chapter 1** | **Probability** | 1 |
| | Random experiments and sample spaces | 1 |
| | Events | 3 |
| | Rules of probability | 7 |
| | Equally likely outcomes | 10 |
| | Permutations and combinations | 13 |
| | Conditional probability and independence | 18 |
| | Bayes' Theorem | 23 |
| | Statistics in action – diagnostic testing in medicine | 28 |
| | Summary | 29 |
| | Review exercise | 31 |
| **Chapter 2** | **Random Variables** | 33 |
| | Review of discrete random variables | 33 |
| | Laws of expectation and variance | 37 |
| | The Binomial distribution | 42 |
| | The Poisson distribution | 46 |
| | Approximating the Binomial with the Poisson | 50 |
| | Review of continuous random variables | 51 |
| | The Normal distribution | 57 |
| | Combining Normal random variables | 61 |
| | Approximating the Binomial with the Normal | 63 |
| | Statistics in action – modelling data | 66 |
| | Summary | 70 |
| | Review exercise | 73 |
| **Chapter 3** | **Sampling** | 75 |
| | Populations and samples | 75 |
| | Simple random sampling | 77 |
| | The distribution of sample means | 79 |
| | The Central Limit Theorem | 82 |
| | A confidence interval for a population mean | 86 |
| | A confidence interval for a population proportion | 90 |
| | Stratified random sampling | 94 |
| | Cluster sampling | 96 |
| | Further sampling methods | 98 |
| | Statistics in action – sampling | 102 |
| | Summary | 106 |
| | Review exercise | 109 |

| | | |
|---|---|---|
| **Chapter 4** | **Hypothesis Testing** | 111 |
| | What is hypothesis testing? | 111 |
| | A test of the population mean | 114 |
| | Assessing the evidence using a *p*-value | 117 |
| | Assessing the evidence using a critical region | 119 |
| | Paired data | 122 |
| | Hypothesis tests and confidence intervals | 124 |
| | Statistics in action – a test of a population proportion | 127 |
| | Summary | 128 |
| | Review exercise | 130 |
| **Preparation for Unit Assessment** | | 131 |
| **Preparation for Course Assessment** | | 133 |
| **Appendices** | | 135 |
| Appendix 1 | Statistical tables | 135 |
| | Table 1 Binomial cumulative distribution function | 135 |
| | Table 2 Poisson cumulative distribution function | 138 |
| | Table 3 Standard Normal cumulative distribution function | 139 |
| | Table 4 Percentage points of the Standard Normal distribution | 140 |
| Appendix 2 | Circles worksheet | 141 |
| Appendix 3 | Data on family size | 142 |
| Appendix 4 | Data on wheat yields | 143 |
| Appendix 5 | A note on estimating the population variance $\sigma^2$ | 145 |
| **Answers** | | 146 |
| **Index** | | 164 |
| **Acknowledgements** | | 166 |

# Preface

This book is part of the *Maths in Action* series, and provides complete coverage of the Statistics 1 unit within the Mathematics and Applied Mathematics courses at Advanced Higher level as detailed in the national course specifications published by the Scottish Qualifications Authority.

Progression to Advanced Higher from Higher Mathematics must take account of the fact that some students will have included the optional Statistics unit at Higher level while others will have chosen the Mathematics 3 unit. A major part of the Statistics (H) and Statistics 1 (AH) units is the study of probability and this book has been carefully written to support students no matter which option they studied at Higher level. Those who did not include Statistics at Higher level will find that no prior knowledge of probability is assumed, while those who have studied Statistics at Higher will find the approach taken retains their interest as it deepens and extends their knowledge.

The content follows the order of the unit specifications, and chapters are numbered to correspond to the unit's learning outcomes. One exception is the section 'Laws of expectation and variance' which, although part of Outcome 1, appears immediately after random variables have been introduced at the beginning of Chapter 2. New topics are introduced through clearly written explanations and worked examples using real data where appropriate. Each section is followed by graded exercises and in some cases these exercises are numbered to distinguish basic work necessary for all students (A exercises) from that of a more demanding nature (B exercises) for those who wish to develop their competence beyond the minimum. Informative answers are provided for every question. Throughout the book, 'Statistics in Action' panels apply and extend the content of each chapter, inviting students to become actively involved in their own learning and to prepare for Outcome 5, the statistical assignment. An end of chapter Summary and Review exercise bring together the work of each chapter, and there are exercises which prepare the student for unit and course assessment.

Calculators with statistical features are widely available. An icon has been used throughout the book to indicate where advice on their use is given and to prompt students to find out what their calculator can do. A short BASIC program has been included so that those with access to a computer may conduct a simulation which demonstrates the Central Limit Theorem.

*Ralph Riddiough*
*John McColl*

# 1 Probability

## *Random experiments and sample spaces*

Probability and statistics are two allied branches of mathematical science. Both are concerned with describing conditions of uncertainty. Here are some examples of situations that might be studied in statistics.

1 When a young person is referred for orthodontic treatment, the consultant rates his/her need for treatment as minimal, mild, moderate or severe. Each young person fits into one and only one of these categories. The consultant does not know before examining a new patient into which category that particular individual will fit.

2 In a class survey, a first-year pupil is chosen at random and asked 'How many brothers and sisters do you have?' Depending on which pupil is chosen from the class, a variety of answers might be obtained: 0, 1, 2, 3, or even more. Since we do not know in advance which pupil will be asked, we do not know in advance which of these answers will be obtained.

3 The game of Yahtzee© is played with five standard dice. This means that each die has six sides, marked respectively with the numbers 1, 2, …, 6. When a standard die is rolled, it is usual to take the score on the die as the number on the face that lands uppermost. At the start of each turn in this game, all five dice are rolled, and the pattern of scores on the five dice is of interest.
A result might be recorded as an ordered quintuple, for example (4, 1, 3, 6, 6), where the first number records the score of one particular die, the second number the score of a particular one of the other dice, and so on.
On my next turn, I might record any of the quintuples $(x_1, x_2, x_3, x_4, x_5)$ where $x_1$, $x_2$, $x_3$, $x_4$, and $x_5$ might each be any of the numbers 1, 2, …, 6.

4 A consumers' organisation obtains three new cars of a particular design and arranges for some drivers to test out their fuel consumption (in miles per gallon, mpg) when driven in motorway conditions at a constant 65 mph. Suppose that the car manufacturer claims that cars of this design give 30 mpg under such conditions. This is likely to be an average figure. Fuel consumption will differ slightly from car to car and from driver to driver, and will also be affected by the precise road conditions met on a test. So, each time one of the cars is tested, the consumers' organisation records some value of fuel consumption, which might lie anywhere in the range 25 mpg to 35 mpg. It would not be known in advance which particular value in that range would actually be recorded.

In statistics, any process by which information is obtained is called an **experiment**. All the above examples describe experiments in a statistical sense, though only the last of them would be considered an experiment from the usual scientific point of view.

The information recorded when an experiment is carried out once is known as the **outcome**. The outcome of a statistical experiment is often a number, for example the surveyed pupil is found to have two brothers or sisters. Sometimes, the outcome is not a number, for example a new orthodontic patient is found to be in severe need of treatment. And, in many important experiments, the outcome is not just one piece of information, for example the scores on five dice turn out to be, in order, (3, 6, 3, 4, 3).

All the above examples are called **random experiments**, because each time the experiment is carried out there are a number of possible outcomes and it is impossible to tell in advance which of them will occur. Statistics is fundamentally concerned with random experiments and seeks to describe the pattern of outcomes that would be seen if the experiment were carried out very often under identical conditions. Each repeat of a random experiment, under specified conditions, is called a **trial**.

The first step in describing or modelling a random experiment is to produce an exhaustive list of all the possible outcomes. In the experiments above, the list would be

$S = \{\text{minimal, mild, moderate, severe}\}$

$S = \{0, 1, 2, 3, \ldots\}$

$S = \{(x_1, x_2, x_3, x_4, x_5): x_1 = 1, 2, 3, 4, 5, 6; \ldots; x_5 = 1, 2, 3, 4, 5, 6\}$

$S = \{x: 25 \leq x \leq 35\}$

The list of all possible outcomes for a random experiment is known as the **sample space *S***. It is extremely important to ensure that the sample space includes every possible outcome of the experiment. No matter how often the experiment is conducted, it must never be possible to record an outcome that is not in the sample space.

In examples 1 and 3 above, we can be sure that the sample space includes just those outcomes that are possible for the experiment. This is not necessarily true in the other examples. It is unlikely that we would ever interview a child with 55 brothers and sisters! Or even 40. But it is not clear where exactly to place an upper limit on the sample space. For this reason, and for mathematical convenience later on, we choose to take as our sample space the set of whole numbers. At least we can be confident that no response will ever lie outside this list.

A similar issue arises in example 4. Initial engineering advice suggests that, no matter how badly one of these cars is driven, the fuel consumption will never get below 25 mpg and, no matter how well driven, the fuel consumption will never get above 35 mpg.

This justifies the use of $S = \{x: 25 \leq x \leq 35\}$. However, to err on the side of caution, it might be better to adopt a larger sample space, such as $S = \{x: 20 \leq x \leq 40\}$ or even $S = \{x: x > 0\}$. The last of these suggestions might be the most convenient mathematically.

## EXERCISE 1.1

**1** Write out suitable sample spaces for each of the following experiments.

**a** A group of 20 people sit the theory paper for their driving test on a particular occasion. The number of them who pass is recorded.

**b** A quality-control inspector examines a fine china ornament produced in a factory, and records that it is either perfect, second quality or imperfect.

**c** Before starting to play tennis against a friend, you agree to play until *either* one of you has won two games more than the other *or* you have completed 30 games. Record the total number of games played.

**d** You toss a coin repeatedly until you get Tails for the first time. Record the total number of tosses required.

**e** A plastic rod is attached to a test rig, where it is placed under strain until it snaps. Record the total time that elapses (in seconds).

**f** You travel to school by the same train every morning for one week (Monday to Friday). Record the number of times the train is late.

**g** An individual is about to start on a weight-reduction diet. Record the change in this person's weight (in kilograms) after one month.

**2** A clerk is responsible for entering 1000 records into a computerised database. Each record consists of values for seven variables (e.g. name, age). Write down a suitable sample space for recording the total number of

**a** values that are entered incorrectly,

**b** records that are entered completely correctly,

**c** values of each of the seven variables that are entered incorrectly.

**3** How many outcomes are there in the sample space for example 3 (page 1)?

## *Events*

We are usually interested in particular collections of outcomes of a random experiment. For example, a guide to the workload of an orthodontics clinic might be the number of patients in the categories moderate or severe. We might want to focus particularly on children with at least two brothers and sisters. In Yahtzee©, we might be interested in obtaining the same score on four of the dice. We might be particularly interested in a car that had fuel efficiency better than 30 mpg. Collections of outcomes like these are subsets of the sample space and called **events**. We could write these events as follows:

$E = \{\text{moderate, severe}\}$

$F = \{2, 3, 4, \ldots\}$

$G = \{(1, 1, 1, 1, 2), (1, 1, 1, 1, 3), (1, 1, 1, 1, 4), (1, 1, 1, 1, 5), (1, 1, 1, 1, 6), (1, 1, 1, 2, 1), (1, 1, 1, 3, 1), (1, 1, 1, 4, 1), (1, 1, 1, 5, 1), (1, 1, 1, 6, 1), \ldots, (1, 6, 6, 6, 6), (2, 6, 6, 6, 6), (3, 6, 6, 6, 6), (4, 6, 6, 6, 6), (5, 6, 6, 6, 6)\}$

$H = \{x\text{: } x > 30\}$

*Example 1* List all the possible events associated with the random experiment of tossing two coins.

*Solution*
The sample space is $S$ = {TT, TH, HT, HH}, where HT (for example) means that the first coin lands with Heads uppermost, while the second coin lands Tails uppermost. The list of all possible events is:

{TT, TH, HT, HH}
{TT, TH, HT}  {TT, TH, HH}  {TT, HT, HH}  {TH, HT, HH}
{TT, TH}  {TT, HT}  {TT, HH}
{TH, HT}  {TH, HH}  {HT, HH}
{TT}  {TH}  {HT}  {HH}
{ }

Notice that the sample space itself is an event, as is every individual outcome. The list of events also includes the **empty set**, { }, which contains none of the outcomes in the sample space and is used to denote something that is impossible (such as getting three Heads when tossing two coins). Including the empty set is very useful mathematically, as we shall see shortly.

Each event in the above list has a meaning, for example:

$S$ = {TT, TH, HT, HH} represents something that is certain to occur
$E$ = {HT, HH} means the first coin lands with Heads uppermost
$F$ = {TT} means both coins land with Tails uppermost.

In some cases, it is easier to find a meaning for an event in terms of what does *not* happen when the event occurs, For example,

$G$ = {TT, TH, HT} means that we do *not* get two Heads

The event $G$ is the opposite of another event, $A$ = {HH}. $G$ consists of all the outcomes in the sample space that do not lie in the event $A$. We say that $G$ is the **complement** of $A$. We write $G = A'$. In other words, the complement of $A$ is the event that occurs when $A$ does not occur. Similarly, $A$ is the complement of $G$, $A = G'$.

Note that $S'$ = { } and { }$'$ = $S$.

We can combine events using the words 'and' and 'or', as illustrated in the following example.

*Example 2* The Advanced Higher Statistics class in a certain school has just four members, Asmat, Beth, Claire and Donald. On a particular day, each may be present (P) or absent (A). The possible attendance entries in the class register can be listed as follows:

$S$ = {PPPP, PPPA, PPAP, PAPP, APPP, PPAA, PAAP, PAPA, APPA, APAP, AAPP, PAAA, APAA, AAPA, AAAP, AAAA}

where, for example, PAPA means that only Asmat and Claire are present.

List the outcomes in the following events:

**a** $E$ = Asmat is present,

**b** $F$ = Donald is present,

**c** $E$ *and* $F$ = Asmat is present *and* Donald is present,

**d** $E$ *or* $F$ = Asmat is present *or* Donald is present,

**e** $G$ = only Claire and Donald are present,

**f** $E$ *and* $G$,

**g** $E$ *and* $E'$,

**h** $E$ *or* $E'$.

*Solution*

**a** $E$ = {PPPP, PPPA, PPAP, PAPP, PPAA, PAAP, PAPA, PAAA}.

**b** $F$ = {PPPP, PPAP, PAPP, APPP, PAAP, APAP, AAPP, AAAP}.

**c** The event $E$ *and* $F$ consists of exactly those outcomes that appear in both $E$ and $F$ separately, i.e. the outcomes that are common to both $E$ and $F$, so

$E$ *and* $F$ = {PPPP, PPAP, PAPP, PAAP}

**d** The event $E$ *or* $F$ consists of those outcomes that appear in either $E$ or $F$ or in both of these events, i.e. every outcome that appears in at least one of the events $E$ or $F$, so

$E$ *or* $F$ = {PPPP, PPPA, PPAP, PAPP, APPP, PPAA, PAAP, PAPA, APAP, AAPP, PAAA, AAAP}

**e** $G$ = {AAPP}.

**f** $E$ *and* $G$ = { }.

$E$ and $G$ have no outcome in common – it is *impossible* for both events to occur on the same day. Pairs of events that have no outcome in common are said to be **disjoint** (or **mutually exclusive**) events.

**g** $E'$ = {APPP, APPA, APAP, AAPP, APAA, AAAP, AAPA, AAAA}. So

$E$ *and* $E'$ = { }.

By definition, $E$ and $E'$ have no outcome in common. They are disjoint. This is true for any event $E$.

**h** $E$ *or* $E' = S$.

By definition, every outcome in $S$ must lie in either $E$ or $E'$. This also is true for any event $E$.

### *Summary*

| **Event** | **Meaning** |
|---|---|
| $S$ | an event that is certain to occur |
| { } | an event that is impossible |
| $E'$ | the event $E$ does not occur |
| $E$ *and* $F$ | both the events $E$ and $F$ occur |
| $E$ *or* $F$ | either $E$ or $F$ occurs or both events occur |
| $E$ *and* $F$ = { } | $E$ and $F$ cannot occur simultaneously |

$E$ *and* $E'$ = { }

$E$ *or* $E' = S$

$E$ *and* $F$ = $F$ *and* $E$

## EXERCISE 1.2

**1** You roll a standard die once, and record the score on the face that lands uppermost.

**a** Write out the sample space for this experiment.

**b** Write out each of the following events as a set:

(i) $A$ = an odd number is recorded,

(ii) $B$ = a prime number (greater than 1) is recorded,

(iii) $C$ = a six is recorded,

(iv) $D$ = a number greater than four is recorded,

(v) $A$ *or* $B$,

(vi) $A$ *and* $B$.

**2** Suppose now that you roll a standard die twice, recording the score on each occasion.

**a** Write out the sample space for this random experiment.

**b** Write out each of the following events as a set:

(i) $E$ = the same score is recorded for both rolls of the die,

(ii) $F$ = the second score is at least three more than the first score,

(iii) $G$ = the first score is twice the second score,

(iv) $H$ = the two scores add up to less than six.

**c** Which pairs of these events are disjoint?

**3** A new treatment for a skin disease is tested on four patients. For each patient, it is recorded whether the treatment is successful (S) or unsuccessful (U).

**a** Write out a suitable sample space for this experiment.

**b** Write out each of the following events as a set:

(i) $A$ = the treatment is successful for exactly three of the patients,

(ii) $B$ = the treatment is successful for at least three of the patients,

(iii) $C$ = the treatment is successful in at most two cases.

**c** Write down an expression for the relationship between the events $B$ and $C$.

**4** The lift in a small hotel connects the ground (G), first (F) and second (S) floors. Three people get into the lift at the ground floor, and the floor where each gets out is recorded.

**a** Write out a suitable sample space for this experiment.
[Hint: can you think why an individual might get out again at the ground floor?]

**b** Write out each of the following events as a set:

(i) no-one gets out at the first floor,

(ii) everyone gets out at the second floor,

(iii) exactly two of the three people get out at the second floor,

(iv) everyone gets out at the same floor.

**5** A Primary 6 class measures the local rainfall (cm) for a one-week period in October.

**a** Write out a suitable sample space for this experiment.

**b** Write out a set corresponding to each of the following events:

(i) rainfall is at least as high as 5 cm, the average for this time of year,

(ii) no rain falls this week,

(iii) rainfall is less than 10 cm,

(iv) rainfall is between 2 cm and 8 cm (inclusive).

**c** Write down, as sets, the complements of the above events.
[Hint: pay particular attention to the different meanings for $<$, $\leq$, $>$ and $\geq$.]

## *Rules of probability*

An event $E$, associated with a particular random experiment, has probability P($E$) which represents how likely it is that $E$ will occur when the experiment is carried out. Probability is measured on a scale from 0 to 1. An event that is impossible has probability 0, and an event that is certain to occur has probability 1.

Suppose that many trials of the random experiment can be carried out. Then we can think of P($E$) as the proportion of trials on which we would expect to observe the event $E$.

For example, suppose we had a 'fair' coin. Then, if we tossed this coin many times, we would expect to obtain roughly the same number of Heads as Tails. In other words we would obtain Heads on roughly $\frac{1}{2}$ of all trials and Tails on the other trials. So the probability of Heads is $\frac{1}{2}$,

$$\mathrm{P(Heads)} = \frac{1}{2}$$

and the probability of Tails is

$$\mathrm{P(Tails)} = \frac{1}{2}$$

An event that is impossible can not be observed, however often we carry out the experiment, so it has probability 0. This means that

$$\mathrm{P}(\{\ \}) = 0$$

On the other hand, an event that is certain to occur must be observed every time we carry out the experiment, so it has probability 1. This means that

$$\mathrm{P}(S) = 1$$

For any event $E$, we know that the proportion of trials on which we observe $E$ must lie between 0 and 1. In symbols,

$$0 \leq \mathrm{P}(E) \leq 1$$

In any experiment, if we can write out the sample space as $S = \{o_1, o_2, o_3, \ldots\}$, then we must observe one and only one of these outcomes on each trial. This means that the proportions of trials on which these outcomes are observed must add up to 1, or

$$\mathrm{P}(o_1) + \mathrm{P}(o_2) + \mathrm{P}(o_3) + \cdots = 1$$

Let us consider the experiment of tossing a 'fair' coin twice. Since, each time we toss the coin, we have P(Heads) = P(Tails) = $\frac{1}{2}$, it follows that each of the possible outcomes TT, TH, HT and HH has probability $\frac{1}{4}$. Notice that P(TT) + P(TH) + P(HT) + P(HH) = 1.

Once we have assigned probabilities to each of the outcomes in a sample space like this, it is an easy matter to find the probability of any event. We can simply add together the probabilities associated with each of the outcomes in the event. For example,

$$\mathrm{P(two\ Heads)} = \mathrm{P(HH)} = \frac{1}{4}$$

$$\mathrm{P(one\ Head)} = \mathrm{P(TH, HT)} = \mathrm{P(TH)} + \mathrm{P(HT)} = \frac{1}{2}$$

$$\mathrm{P(no\ Head)} = \mathrm{P(TT)} = \frac{1}{4}$$

*Example* In a certain university, first-year students who are registered to study in the mathematical science area must take exactly three different subjects, at least one of which must be mathematics (M), statistics (S), or computing (C). Letting 'A' represent any other subject, a suitable sample space for recording each of these students' curriculum is:

$$S = \{\text{MSC, MSA, MCA, SCA, MAA, SAA, CAA}\}$$

The proportions of students in these seven subject groups are (respectively) 0.50, 0.10, 0.20, 0.05, 0.05, 0.05, 0.05. [These proportions add up to 1.]

Suppose now that a student is chosen at random from this group of students, to take part in a survey of student opinion. Find the probability of the following events:

**a** $E$ = the student is studying mathematics,
**b** $F$ = the student is not studying mathematics,
**c** $G$ = the student is studying computing,
**d** $H$ = the student is studying computing but not mathematics,
**e** $E$ *or* $H$,
**f** $E$ *or* $G$,
**g** $E$ *and* $G$.

*Solution*

**a**
$$E = \{\text{MSC, MSA, MCA, MAA}\}$$
$$\text{P}(E) = \text{P(MSC)} + \text{P(MSA)} + \text{P(MCA)} + \text{P(MAA)}$$
$$= 0.50 + 0.10 + 0.20 + 0.05 = 0.85$$

In other words, 85% of this population of students study mathematics.

**b**
$$F = E' = \{\text{SCA, SAA, CAA}\}$$
$$\text{P}(F) = \text{P(SCA)} + \text{P(SAA)} + \text{P(CAA)}$$
$$= 0.05 + 0.05 + 0.05 = 0.15$$

In other words, 15% of this student population do not study mathematics.
Notice that $\text{P}(E') = 1 - \text{P}(E)$, or $\text{P}(E) + \text{P}(E') = 1$.
Writing this out in full,

$$\text{P}(E) + \text{P}(E') = [\text{P(MSC)} + \text{P(MSA)} + \text{P(MCA)} + \text{P(MAA)}] + [\text{P(SCA)} + \text{P(SAA)} + \text{P(CAA)}]$$

which is the sum of the probabilities of all the outcomes in the sample space, which we have seen must add to 1.
This is because $E'$ consists of exactly those outcomes that are not in $E$.

**c**
$$G = \{\text{MSC, MCA, SCA, CAA}\}$$
$$\text{P}(G) = 0.50 + 0.20 + 0.05 + 0.05 = 0.80$$

**d**
$$H = \{\text{SCA, CAA}\}$$
$$\text{P}(H) = 0.05 + 0.05 = 0.10$$

**e**
$$E \text{ or } H = \{\text{MSC, MSA, MCA, MAA, SCA, CAA}\}$$
$$\text{P}(E \text{ or } H) = 0.50 + 0.10 + 0.20 + 0.05 + 0.05 + 0.05 = 0.95$$

We can write this out as

$$\begin{aligned}\text{P}(E \text{ or } H) &= \text{P(MSC)} + \text{P(MSA)} + \text{P(MCA)} + \text{P(MAA)} + \text{P(SCA)} + \text{P(CAA)} \\ &= [\text{P(MSC)} + \text{P(MSA)} + \text{P(MCA)} + \text{P(MAA)}] + [\text{P(SCA)} + \text{P(CAA)}] \\ &= \text{P}(E) + \text{P}(H)\end{aligned}$$

This works out because $E$ and $H$ are disjoint events, so the outcomes in the list for $E$ or $H$ belong to one and only one of the lists for $E$ and $H$ separately. This result is sometimes called **the addition rule for disjoint events**.

**f** $E$ *or* $G$ = {MSC, MSA, MCA, SCA, MAA, CAA}

P($E$ *or* $G$) = 0.50 + 0.10 + 0.20 + 0.05 + 0.05 + 0.05 = 0.95

Now, P($E$ *or* $G$) is not equal to P($E$) + P($G$), since $E$ and $G$ are not disjoint events.

**g** $E$ *and* $G$ = {MSC, MCA}

P($E$ *and* $G$) = 0.50 + 0.20 = 0.70

Notice that

P($E$ *or* $G$) + P($E$ *and* $G$) = 1.65 = P($E$) + P($G$).

The result P($E$ *or* $G$) = P($E$) + P($G$) – P($E$ *and* $G$) is sometimes called **the general addition rule**.

The addition rule for disjoint events can be extended as follows. Suppose that $E_1$, $E_2$, $E_3$, … are disjoint events. In other words, $E_i$ *and* $E_j$ = { } whenever $i$ is not equal to $j$. Then,

$$P(E_1 \text{ or } E_2 \text{ or } E_3 \text{ or } \ldots) = P(E_1) + P(E_2) + P(E_3) + \cdots$$

### *Summary*

For any event $E$, $0 \le P(E) \le 1$.

P($S$) = 1.

P({ }) = 0.

P($E'$) = 1 – P($E$).

For disjoint events $E$ and $F$, P($E$ *or* $F$) = P($E$) + P($F$).

For disjoint events $E$ and $F$, P($E$ *and* $F$) = P({ }) = 0.

For any pair of events $E$ and $F$, P($E$ *or* $F$) = P($E$) + P($F$) – P($E$ *and* $F$).

## EXERCISE 1.3

**1** Refer back to question 1 of Exercise 1.2. Assume that every possible outcome of rolling the die has probability $\frac{1}{6}$.

**a** Find the probability of each of the events $A$, $B$, $C$ and $D$.

**b** Find P($A$ *or* $B$) and P($A$ *and* $B$). Verify that the general addition rule holds for the events $A$ and $B$ in this example.

**2** Refer back to question 2 of Exercise 1.2. Assume that every possible outcome in the sample space has probability $\frac{1}{36}$. Find the probability of each of the events $E$, $F$, $G$ and $H$.

**3** Houses in a local area may have two free newspapers delivered, but delivery is patchy. A survey suggests that 70% of the households get paper X and 80% get paper Y. 60% of all the households get both papers delivered.

**a** What proportion of households get at least one of the papers delivered?

**b** What proportion of households get neither paper delivered?

**4** I sometimes call in unexpectedly to visit an old couple I know. I have noticed that Elsie is in only 50% of the time while Frank is at home 65% of the time. On 10% of the occasions I call, I get no reply because both are out. On what percentage of times that I call do I get to see both Elsie and Frank?

**5** Show that the general addition rule reduces to the addition rule for disjoint events when $E$ and $F$ are disjoint.

## *Equally likely outcomes*

We can find out the probability of an event in two main ways. The first is to carry out the experiment very many times, and use the results of these trials to estimate the probabilities. For example, drawing pins of a certain type were tossed 5000 times. On 2778 occasions, the drawing pins landed with their flat head on the ground ('pointing up') and on the other 2222 occasions the drawing pins landed 'pointing down'. As a result, we could estimate

P(pin lands 'pointing up') by $\frac{2778}{5000} = 0.5556$

and

P(pin lands 'pointing down') by $\frac{2222}{5000} = 0.4444$

The second main way to determine probabilities only works with certain kinds of experiments. It is to think about the conditions of the random experiment and realise that there are patterns to the possible outcomes that allow us to assign probabilities to them without carrying out a lot of trials. The simplest experiments of this type are the ones that have **equally likely outcomes**. Here are some examples.

1. Tossing a 'fair' coin. Here we assume that Heads and Tails are equally likely. In other words, in the long run the two outcomes arise equally often. So, each outcome must occur roughly half the times the experiment is carried out. This suggests that $P(\text{Heads}) = P(\text{Tails}) = \frac{1}{2}$.
2. Rolling a 'fair' die. Here all six faces are equally likely. So
   $P(1) = P(2) = \cdots = P(6) = \frac{1}{6}$.
3. A lottery. A bank trying to find out about its customers posts a questionnaire to 10 000 of them (chosen from its computerised records). Each of the customers who replies is entered into a prize draw, with one prize of £1000. If $N$ of the customers reply, then each of them is equally likely to win the prize. So the probability that a particular customer (who replies) wins is $\frac{1}{N}$.

In each of these examples, an 'equally likely outcomes' model appears appropriate, based on the conditions of the experiment. Such models are capable of a good deal of sophistication.

Suppose we toss a fair coin three times. The possible outcomes of this experiment are

either H or T *followed by* either H or T *followed by* either H or T

Each possible outcome of the first toss can be followed by every possible outcome of the second toss, and so on, so there are $2 \times 2 \times 2 = 8$ possible (ordered) outcomes for the entire experiment. We may write these out as:

HHH HHT HTH HTT THH THT TTH TTT

Again, all of these eight outcomes are equally likely, so each has probability $\frac{1}{8}$. This means that:

$$P(0 \text{ Heads}) = P(TTT) = \frac{1}{8}$$

$$P(\text{exactly 1 Head}) = P(HTT) + P(THT) + P(TTH) = \frac{3}{8}$$

$$P(\text{exactly 2 Heads}) = P(HHT) + P(HTH) + P(THH) = \frac{3}{8}$$

$$P(3 \text{ Heads}) = P(HHH) = \frac{1}{8}$$

From the above examples, we can obtain the following general result.

In an 'equally likely outcomes' model, the probability of any event $E$ is

$$P(E) = \frac{\text{number of different outcomes in } E}{\text{number of different outcomes in } S}$$

*Example* Suppose we roll a blue die and a red die together, recording the score on the blue die first. Since there are six possible outcomes for each die, then there are $36 = 6 \times 6$ possible outcomes of the entire experiment. These are all the pairs $(b, r)$ where $b$ and $r$ may each take the values 1, 2, ..., 6. Assuming both dice are fair, find the probabilities of the events:

**a** $F$ = the total score on the dice is 5,
**b** $E_1$ = a 6 is scored on both dice,
**c** $E_2$ = a 6 is scored on neither die,
**d** $E_3$ = a 6 is scored on exactly one of the dice.

*Solution*

**a** If the dice are fair, then all 36 outcomes are equally likely. So
$P(F) = P((1, 4)) + P((2, 3)) + P((3, 2)) + P((4, 1)) = \frac{4}{36} = \frac{1}{9}$

**b** $P(E_1) = P((6, 6)) = \frac{1}{36}$

**c** For the event $E_2$, we require $b = 1, 2, 3, 4$ or $5$ and $r = 1, 2, 3, 4$ or $5$. So, the number of (ordered) outcomes in $E_2$ is $5 \times 5 = 25$. Hence,

$$P(E_2) = \frac{25}{36}$$

**d** Of the 36 outcomes, we have already found that one lies in $E_1$ and 25 others lie in $E_2$. All the remaining outcomes must lie in $E_3$, since every outcome must lie in either $E_1$ or $E_2$ or $E_3$. This means there are $36 - 1 - 25 = 10$ outcomes in $E_3$. So

$$P(E_3) = \frac{10}{36}$$

## EXERCISE 1.4

**1** A fair coin is tossed four times.

**a** How many possible (ordered) outcomes are there to this experiment? Write them all out.

**b** Find the probability of obtaining:

**(i)** Heads at least once,

**(ii)** Heads and Tails equally often,

**(iii)** Tails at least three times.

**2** In a multiple-choice examination, five possible answers are given to each question. Only one of the answers for each question is correct.

**a** Joe has to guess the answer to one of the questions. Find the probability that he guesses the correct answer.

**b** Sue has to guess the answer to two questions. How many possible (ordered) outcomes are there for her two guesses? What is the probability that she guesses

**(i)** both answers correctly,

**(ii)** neither answer correctly,

**(iii)** exactly one of the answers correctly?

**3** **a** A roulette wheel has 38 pockets, which are numbered respectively 1, 2, ..., 36, 0 and 00. When the wheel is spun, a ball goes round and round, eventually dropping into one of the pockets. The winning number is the number of the pocket into which the ball drops. Assuming that 0 and 00 are neither odd nor even, find the probability that the winning number is

**(i)** odd, **(ii)** even,

**(iii)** either 0 or 00, **(iv)** between 30 and 36 (inclusive).

**b** Suppose that the roulette wheel is spun twice in succession.

**(i)** How many possible (ordered) outcomes are there?

**(ii)** Find the probability that the winning number both times is either the 0 or the 00.

**4** The faces of a regular ten-sided spinner are marked with the digits 0, 1, ..., 9.

**a** It is spun once, and the score $X_1$ is recorded. Noting that 0 is divisible by any whole number, find the probability of the events

**(i)** $E =$ '$X_1$ is divisible by 2',

**(ii)** $F =$ '$X_1$ is divisible by 3',

**(iii)** *E and F*,

**(iv)** *E or F*.

**b** The spinner is spun a second time, and the score $X_2$ is recorded. The outcome of the two spins is taken as $Y = 10X_1 + X_2$. Find the probability of the events

**(i)** $G =$ '$Y$ is divisible by 2',

**(ii)** $H =$ '$Y$ is divisible by 5',

**(iii)** $M =$ '$Y$ is not divisible by 10',

**(iv)** *G and H*,

**(v)** *G and M*,

**(vi)** *H and M*.

**c** Write down an expression that relates the events $G$, $H$ and $M$.

## *Permutations and combinations*

The formula

$$P(E) = \frac{\text{number of different outcomes in } E}{\text{number of different outcomes in } S}$$

means that calculating probabilities within an equally likely outcomes model boils down to counting the number of different outcomes associated with various events. We shall now look at some special ways of counting outcomes without having to write them all down.

Suppose that the senior school's boys 1500 m race is to be contested by eight athletes. In how many different possible ways can the Gold, Silver and Bronze medals be distributed?

- There are eight possible winners of the Gold medal.
- For each possible winner of the Gold medal, there are seven possible winners of the Silver medal (since the samc athlete cannot win both medals).
- For each possible pair of winners of the Gold and Silver medals, there are six possible winners of the Bronze medal.

In total, then, there are

$$8 \times 7 \times 6 = 336$$

different ways in which the three medals may be distributed. (Of course, when the race is actually run, the medals will be distributed in just one of these possible ways.)

So, there are 336 ordered choices of three athletes from a field of eight. Every *ordered* choice of $r$ from $n$ objects (where $r = 1, \ldots, n$) is known as a **permutation**. There are 336 different permutations of three from eight distinct objects (athletes). In general, when making an ordered choice of $r$ from $n$ objects, there are:

- $n$ ways to choose the first object,
- $n - 1$ ways to choose the second object (for each possible choice of the first),
- $n - 2$ ways to choose the third object (for each possible choice of the first two),

  ⋮
- $n - r + 1$ ways to choose the $r$th object (for each possible choice of the first $r - 1$).

So the total number of different possible permutations is

$${}^nP_r = n \times (n-1) \times (n-2) \times \ldots \times (n+1-r), \quad r = 1, 2, \ldots, n$$

It is useful to adopt the following notation. For any positive integer, $n$,

$$n! = n \times (n-1) \times (n-2) \times \ldots \times 1 \quad (n! \text{ is pronounced '}n\text{ factorial'})$$

By convention $0! = 1$.

So

$${}^nP_r = n(n-1)(n-2)\ldots(n+1-r) = \frac{n(n-1)(n-2)\ldots(n+1-r)(n-r)\ldots 1}{(n-r)\ldots 1}$$

$$= \frac{n!}{(n-r)!}$$

When $n = 8$, $r = 3$,

$$n! = 8! = 8 \times 7 \times 6 \times 5 \times 4 \times 3 \times 2 \times 1 = 40\,320$$

$$(n - r)! = 5! = 5 \times 4 \times 3 \times 2 \times 1 = 120$$

$${}^nP_r = \frac{n!}{(n-r)!} = \frac{40\,320}{120} = 336 \text{ (as before)}$$

It should be obvious from the definition that ${}^nP_r$ must always be a non-negative integer.

In the above example, suppose that the school's running track has just six lanes. The race will be run in two heats, each of four athletes. (The best two athletes from each heat will then contest the final.) In how many different ways can four athletes be chosen for the first heat?

At first sight, this seems similar to the previous question. So, we start by calculating that there are

$$8 \times 7 \times 6 \times 5$$

permutations of four from eight athletes. But in this case, it does not matter which order the athletes are chosen in to make up the first heat. For example, athletes A, B, C and D would still compete against each other in the first heat if they were chosen in any of the following orders:

| | | | | | |
|---|---|---|---|---|---|
| ABCD | ABDC | ACBD | ACDB | ADBC | ADCB |
| BACD | BADC | BCAD | BCDA | BDAC | BDCA |
| CABD | CADB | CBAD | CBDA | CDAB | CDBA |
| DABC | DACB | DBAC | DBCA | DCAB | DCBA |

An *unordered* choice of $r$ from $n$ objects ($r = 0, 1, \ldots, n$) is known as a **combination**. The combination A, B, C, D corresponds to the 24 different permutations listed above. In fact, every combination of four athletes corresponds to 24 permutations, in a similar way, so the number of distinct combinations must be

$$\frac{8 \times 7 \times 6 \times 5}{24} = 70$$

There are, then, 70 distinct (unordered) ways of choosing four of the eight athletes to make up the first heat of this race.

In this example, each combination corresponded to 24 permutations. 24 is the number of ways in which any four athletes could be permuted among themselves:

$$4 \times 3 \times 2 \times 1 = 4! = 24$$

In general, any combination of $r$ from $n$ ($r = 1, 2, \ldots, n$) objects corresponds to $r!$ permutations. Hence the number of distinct combinations is

$$\binom{n}{r} = \frac{{}^nP_r}{r!} = \frac{n!}{r!(n-r)!}$$

Conventionally, for $r = 0$, $\binom{n}{0} = 1$, i.e. there is just one way of choosing 0 from $n$ objects.

In the above example, $n = 8$ and $r = 4$, so

$$\binom{n}{r} = \binom{8}{4} = \frac{8!}{4!4!} = \frac{8 \times 7 \times \not{6} \times 5 \times \not{4} \times \not{3} \times \not{2} \times \not{1}}{(4 \times \not{3} \times \not{2} \times 1)(\not{4} \times \not{3} \times \not{2} \times \not{1})} = 70 \text{ (as before)}$$

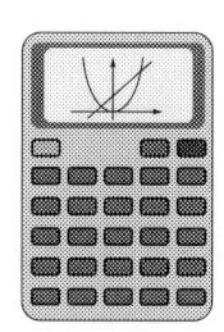

Most scientific and graphic calculators have a button or menu item for calculating the number of permutations and combinations. Look for $n!$, $nPr$ and $nCr$. $^nC_r$ is another popular notation for $\binom{n}{r}$.
Find out what your calculator can do.

*Example* In a class of 30 pupils, what is the probability that at least two pupils share the same birthday?

*Solution*
Assume that none of the pupils was born on 29 February. Assume also that, in the population as a whole, equal numbers of children are born on each of the other 365 days of the year. Considering the 30 children in order (say, in the order of the class register), the number of possible sets of birthdays is

$$365 \times 365 \times 365 \times \ldots \times 365 = 365^{30}$$

With the above assumptions, all of these possibilities are equally likely.

The number of ways in which all the children can have different birthdays is:

$$365 \times 364 \times \ldots \times 336$$

So

$$\text{P(all the children have different birthdays)} = \frac{365 \times 364 \times \ldots \times 336}{365^{30}}$$

The event we are interested in is the complementary event to this, so

$$\text{P(at least two children share the same birthday)} = 1 - \frac{365 \times 364 \times \ldots \times 336}{365^{30}}$$
$$= 0.706$$

This might seem surprisingly large. In fact, the probability of at least two children sharing a birthday is greater than $\frac{1}{2}$ whenever the size of the group is at least 23.

## EXERCISE 1.5A

**1** Evaluate the following expressions (**i**) with and (**ii**) without the aid of the $n!$ button on your calculator.

**a** 5! **b** 6! **c** 10! **d** 20!

**2** Evaluate the following expressions (**i**) with and (**ii**) without the aid of the $nPr$ button on your calculator.

**a** $^4P_3$ **b** $^{10}P_5$ **c** $^7P_2$ **d** $^{12}P_4$ **e** $^8P_6$

**3** Evaluate the following expressions (**i**) with and (**ii**) without the aid of the $nCr$ button on your calculator.

**a** $\binom{4}{3}$ **b** $\binom{10}{5}$ **c** $\binom{7}{2}$ **d** $\binom{12}{4}$ **e** $\binom{8}{6}$

4 A sales representative is planning a business trip during which she has to visit clients in Arbroath, Brechin, Crieff, Dunfermline and Elgin. In how many different ways could she order her visits?

5 A truck driver has a choice of three routes from town A to town B, two routes from town B to town C and four routes from town C to town D. If he has to drive a load from town A to town D, via towns B and C, how many different routes can he take for his journey?

6 In a wine-tasting evening, each taster is required to rank three wines X, Y and Z in order of preference.
  **a** In how many different ways can this be done? Write them all out.
  **b** Assuming that the people taking part in this event have no ability to judge wine and so guess the order in which to place the three wines, what is the probability of one of them ranking X as **(i)** first, **(ii)** second, **(iii)** third?

7 1000 raffle tickets have been sold for school funds. There will be five prizes of equal value. Each ticket may win at most one prize.
  **a** Write down an expression for the number of different possible combinations of winning tickets.
  **b** Suppose you have bought three tickets. Find the probability that you will win at least one prize.
  [Hint: first find the probability of the complementary event.]

## EXERCISE 1.5B

1 **a** Evaluate ${}^5P_r$ for $r = 1, 2, 3, 4, 5$.
  **b** In general, what is ${}^nP_n$?
  **c** In general, what is the relationship between ${}^nP_n$ and ${}^nP_{n-1}$?

2 **a** Evaluate $\binom{n}{r}$ for $r = 0, 1, \ldots, n$ when **(i)** $n = 3$, **(ii)** $n = 4$, **(iii)** $n = 5$.
  **b** In general, find: **(i)** $\binom{n}{0}$ **(ii)** $\binom{n}{1}$ **(iii)** $\binom{n}{n-1}$ **(iv)** $\binom{n}{n}$
  **c** In general, what is the relationship between $\binom{n}{r}$ and $\binom{n}{n-r}$?
  **d** In general, what do you think might be the value of $\binom{n}{0} + \binom{n}{1} + \cdots + \binom{n}{n}$?

3 In a multiple-choice examination, there are ten questions. Five possible answers are given for each question, only one of which is correct.
  **a** In how many different ways could the examination be answered?
  **b** Find the probability that a student who guesses the answer to every question gets them all wrong.

4 a In the game of Yahtzee© described on page 1, find the number of outcomes in the sample space describing the outcome of the first roll of the five dice.
b Hence find the probability of obtaining the same score on all five dice.
c Find the probability of obtaining a 1 on exactly four dice.
d Hence find the probability of obtaining the same score on exactly four dice.
e Hence find the probability of obtaining the same score on at least four dice.

5 An electrical engineer supervises staff who maintain telecommunications equipment at 25 remote sites. In each of the next two weeks, he intends to visit five different sites, chosen at random so that the staff do not know in advance of his visits.
a In how many different ways can he choose five sites to visit next week?
b The engineer will not visit any site two weeks running. So in how many different ways can he now choose five sites to visit in the second week?
c He has chosen to visit sites A, B, C, D and E next week. He will visit one of the sites each day. In how many different ways can he arrange his visits?
d Assuming that he arranges the order of next week's visits at random, find the probability that he visits site A on Monday.

6 Five representatives are to be chosen from a sixth year of 60 girls and 40 boys.
a How many different groups of five representatives can be chosen?
b How many different groups of three girls could be chosen from this sixth year?
c How many different groups of two boys could be chosen from this sixth year?
d If the five representatives are chosen at random, show that the probability that exactly three of them are girls (and hence exactly two are boys) is

$$\frac{\binom{60}{3}\binom{40}{2}}{\binom{100}{5}}$$

e Write down a general expression for the probability that a random selection of five representatives includes exactly $x$ girls, for $x = 0, 1, \ldots, 5$.

## *Conditional probability and independence*

*Example 1* An international banking group employs a large number of university graduates each year. In a typical year, the bank receives applications from 6000 prospective graduates. It calls for interview just 900 of those who apply. Of those interviewed, 300 are invited back for a second interview, and 100 of those who are interviewed a second time are offered a job with the bank. Suppose you are a prospective graduate who applies for a job with this bank in a typical year. Find

**a** the probability that you are offered a job,

**b** the probability that you are offered a job given that you are invited to come for an interview,

**c** the probability that you are offered a job given that you are invited back for a second interview.

*Solution*

**a** 100 out of the 6000 applicants will be offered a job, so:

$$\text{P(you are offered a job)} = \frac{100}{6000} = \frac{1}{60} = 0.0167$$

**b** Just 900 applicants are being interviewed, so:

$$\text{P(you are offered a job, given that you are interviewed)} = \frac{100}{900} = \frac{1}{9} = 0.1111$$

**c** Just 300 applicants are interviewed a second time, so:

$$\text{P(you are offered a job, given that you are interviewed a second time)}$$
$$= \frac{100}{300} = \frac{1}{3} = 0.3333$$

Why is the probability changing in this example? It is because you are gathering more information at each stage. A probability calculated on the basis of further knowledge (or further assumptions) is called a **conditional probability**. In this example, we have found that the *conditional probability* of being made a job offer *given that* you have been offered a (first) interview is $\frac{1}{9}$. The *conditional probability* of being offered a job *given that* you have been called for a second interview is $\frac{1}{3}$. Both of these probabilities contrast with the original, **unconditional probability** of $\frac{1}{60}$.

Suppose now that you intend to roll a fair die twice in succession, recording the score on the uppermost face each time. Let $E$ be the event 'total score is 6'. There are 36 different outcomes in the sample space for this experiment. There are five different outcomes in the event $E$, namely (5, 1), (4, 2), (3, 3), (2, 4), (1, 5). So $P(E) = \frac{5}{36}$.

Now suppose that the first roll gives a score of 3. Let the event $F$ be 'first score is 3'. There are now only six possible outcomes in total: (3, 1), (3, 2), (3, 3), (3, 4), (3, 5), (3, 6). Of these, just (3, 3) is associated with the event $E$. So, the *conditional probability* of $E$ *given* $F$ is

$$P(E|F) = \frac{1}{6}$$

In general, when $P(F) > 0$, we can obtain a conditional probability from unconditional probabilities using the following formula

$$P(E|F) = \frac{P(E \text{ and } F)}{P(F)}$$

In this example,

$$P(E \text{ and } F) = P(\text{total score is 6 } and \text{ first score is 3}) = P((3, 3)) = \frac{1}{36}$$

$$P(F) = P(\text{first score is 3}) = \frac{1}{6}$$

so

$$P(E|F) = \frac{\frac{1}{36}}{\frac{1}{6}} = \frac{1}{6} \text{ (as before)}$$

Note in passing that the same rules that are used to manipulate unconditional probabilities are used with conditional probabilities. For example,

$$P(E'|F) = 1 - P(E|F)$$

Now consider the event $G$ = 'total score is 7'.

$$P(G) = P((6, 1)) + P((5, 2)) + P((4, 3)) + P((3, 4)) + P((2, 5)) + P((1, 6)) = \frac{6}{36} = \frac{1}{6}$$

$$P(G \text{ and } F) = P((3, 4)) = \frac{1}{36}$$

$$P(G|F) = \frac{P(G \text{ and } F)}{P(F)} = \frac{\frac{1}{36}}{\frac{1}{6}} = \frac{1}{6} = P(G)$$

Since the unconditional and conditional probabilities of $G$ are the same, this means that knowledge about $F$ does not affect the probability of $G$. We say that $G$ and $F$ are **independent** events.

To allow for situations where either event has probability 0, two events $E$ and $F$ are formally defined to be independent if and only if

$$P(E \text{ and } F) = P(E) \times P(F)$$

This is sometimes called **the multiplication rule for independent events**.

When $P(E)$ and $P(F)$ are both greater than 0, this definition implies that

$$P(E|F) = \frac{P(E \text{ and } F)}{P(F)} = \frac{P(E) \times P(F)}{P(F)} = P(E)$$

and

$$P(F|E) = \frac{P(F \text{ and } E)}{P(E)} = \frac{P(E \text{ and } F)}{P(E)} = \frac{P(E) \times P(F)}{P(E)} = P(F)$$

The multiplication rule for independent events can be extended to any number of independent events. If $E_1$, $E_2$, $E_3$, ... are independent events, then

$$P(E_1 \text{ and } E_2 \text{ and } E_3 \text{ and } \ldots) = P(E_1) \times P(E_2) \times P(E_3) \times \ldots$$

*Example 2* A children's board game is played with a deck of 40 cards. Ten cards are coloured blue, red, yellow and green. One card of each colour is numbered 1, 2, ..., 10. One of the cards is drawn from this pack. Without replacing the first card, a second card is drawn from the pack. Let $E$ = 'first card is yellow', $F$ = 'first card is a 9', $G$ = 'second card is yellow'.

**a** Show that $E$ and $F$ are independent events.

**b** Find $P(E \textit{ and } G)$.

*Solution*

**a** $P(E) = \frac{10}{40} = \frac{1}{4}$, $P(F) = \frac{4}{40} = \frac{1}{10}$

$P(E \textit{ and } F) = P$ (first card is the yellow 9) $= \frac{1}{40}$

so $P(E \textit{ and } F) = P(E) \times P(F)$

i.e. the events $E$ and $F$ are independent.

**b** The definition of conditional probability states that

$$P(G|E) = \frac{P(G \textit{ and } E)}{P(E)} = \frac{P(E \textit{ and } G)}{P(E)}$$

i.e. $P(E \textit{ and } G) = P(G|E) \times P(E)$

The result is sometimes known as the **general multiplication rule for probability**.

We already know $P(E)$. To use the above result, we need to find $P(G|E)$. Given that $E$ has occurred, then one yellow card has been removed from the pack, leaving 39 cards of which 9 are yellow. So

$$P(G|E) = \frac{9}{39}$$

and

$$P(E \textit{ and } G) = P(G|E) \times P(E) = \frac{9}{39} \times \frac{1}{4} = \frac{3}{52}$$

## Summary

For any events $E$ and $F$, with $P(F) > 0$,

$$P(E|F) = \frac{P(E \textit{ and } F)}{P(F)}$$

For any events $E$ and $F$, with $P(F) > 0$,

$$P(E \textit{ and } F) = P(E|F)P(F)$$

$E$ and $F$ are independent when $P(E \textit{ and } F) = P(E) \times P(F)$.

If $E_1$, $E_2$, $E_3$, ... are independent events, then

$$P(E_1 \textit{ and } E_2 \textit{ and } E_3 \textit{ and } \ldots) = P(E_1) \times P(E_2) \times P(E_3) \times \ldots$$

## EXERCISE 1.6A

1 A bag contains 20 marbles, which are identical apart from colour. Twelve of the marbles are red and the remaining eight are yellow. Without looking, you choose one marble from the bag, note its colour, but do not return it to the bag. You then choose a second marble from the bag, again without looking. Let $E$ = 'first marble is yellow', and $F$ = 'second marble is yellow'. Find

**a** $P(F|E)$ **b** $P(F|E')$ **c** P($E$ *and* $F$)

2 Two bags each contain ten red marbles and ten yellow marbles. Without looking, you choose one marble from the first bag and place it in the second bag. Again without looking, you then choose a marble out of the second bag. Let $E$ = 'first marble is yellow' and $F$ = 'second marble is yellow'. Find

**a** $P(F|E)$ **b** $P(F|E')$ **c** P(both marbles are red)

3 The students in a second-year university class were asked about their consumption of alcohol. Some students did not drink alcohol at all. The weekly consumption of alcohol by the other students was compared with the most recent Government guidelines for women and men. Students whose consumption of alcohol was above these guidelines were classed as heavy drinkers, the others were classed as moderate drinkers. The table below shows the results separately for male and female students.

| | Non-drinker | Moderate drinker | Heavy drinker | Total |
|---|---|---|---|---|
| Female students | 15 | 128 | 26 | 169 |
| Male students | 2 | 32 | 18 | 52 |
| Total | 17 | 160 | 44 | 221 |

A student is picked at random from this class. Find the probability that:

**a** the student is female,
**b** the student is a non-drinker,
**c** the student is a heavy drinker,
**d** the student is a non-drinker *given* that the student is female,
**e** the student is a heavy drinker *given* that the student is female,
**f** the student is female *given* that the student is a non-drinker,
**g** the student is female *given* that the student is a heavy drinker.

4 A fair coin is tossed repeatedly until it lands Heads up. Assuming that the outcomes of different tosses are independent, find the probability that the coin first lands Heads up after

**a** one toss, **b** two tosses.

5 A light aircraft has two engines. Independently of the other, each of these engines has probability 0.01 of failing during a flight. Find the probability that:

**a** both engines fail during the same flight,
**b** neither engine fails during a flight,
**c** exactly one engine fails during a flight.

6 An insurance claims adjuster has to get approval from her Section Head (SH) and then from their Manager (M) before settling a claim. She believes that her Section Head will approve 80% of her proposals for settlement, and that their Manager will subsequently approve 90% of her proposals that have been approved by her Section Head. What proportion of her proposals for settlement are approved by both of them?

7 Suppose that $E$ and $F$ are events in a sample space, such that $P(E) = 0.6$ and $P(F) = 0.3$. Use the general addition rule to find $P(E \text{ or } F)$ when:
**a** $E$ and $F$ are disjoint, **b** $E$ and $F$ are independent, **c** $P(E|F) = 0.8$.

8 Three friends are playing table tennis at their local sports centre. In a singles game, Ambreen has probability $\frac{3}{4}$ of beating Barry and probability $\frac{2}{5}$ of beating Carmen. Barry has probability $\frac{1}{3}$ of beating Carmen. (It is impossible to draw a table tennis game.) On this occasion, Ambreen will play Barry in the first game, and the winner will play Carmen. Find the probability that
**a** Ambreen wins both these games,
**b** Barry wins both these games.

9 Sweet peas of a certain kind have either purple or red flowers. The colour is determined by a particular gene which has two forms, $C$ and $c$. Every plant has two of these genes, so plants may have genotype $CC$, $Cc$ or $cc$. Only plants with genotype $cc$ have red flowers. An offspring obtains one of its genes from each parent and the genes inherited from the two parents are determined independently. A parent plant is equally likely to donate each of its genes to an offspring. (So, for example, a $Cc$ parent is equally likely to donate a $C$ or a $c$ gene to one of its offspring.) For every possible combination of parental genotypes, find the probability that an offspring has red flowers.

## EXERCISE 1.6B

1 A certain board game is played with a standard die. Your turn consists of rolling the die once, but whenever you throw a 6 you get to roll again. Your score for your turn is the total score from all the rolls of the die. Find:
**a** the probability of scoring 5,
**b** the probability of scoring 6,
**c** the probability of scoring at least 7,
**d** the probability of scoring 10,
**e** the probability of scoring at least 13.

2 Suppose that the events $E$ and $F$ are disjoint, with $P(E) > 0$ and $P(F) > 0$. Show that $E$ and $F$ can not also be independent.

3 A tutorial group has ten student members. Eight of the students always attend tutorials, but the other two sometimes miss them. Student X attends 80% of tutorials, while student Y attends 60% of tutorials.
**a** Assuming that students X and Y attend or do not attend tutorials independently of one another, find the probability that the next tutorial is attended by a total of **(i)** ten students, **(ii)** eight students.
**b** Hence find the probability that the tutorial is attended by nine students.

4 A study was recently carried out in Scotland to improve teenagers' ability to manage their own asthma. The study depended on recruiting teenagers with asthma at the rate of four each week. Each Sunday evening, four teenagers who had previously expressed an interest in participating in the study were telephoned to remind them of their initial appointment the next week. It was thought that a teenager misses an appointment, after a reminder like this, with probability $\frac{1}{10}$.

a Write out all the (ordered) possible outcomes for the four teenagers in a given week.

b Assuming that the teenagers all act independently, write down the probability of each possible outcome in the sample space.

c If $X$ is the number of teenagers who miss their appointments, in a randomly selected week, find the probability that $X = 0, 1, 2, 3, 4$.

## Bayes' Theorem

If $E$ is any event in a sample space $S$, then we have already seen that

$$E \text{ and } E' = \{\ \} \quad \text{and} \quad E \text{ or } E' = S.$$

$E$ and $E'$ are said to **partition** the sample space. Every outcome in $S$ is in one and only one of the events $E$ and $E'$.

In general, events $E_1, E_2, \ldots$ with non-zero probability are said to partition the sample space if every pair of these events is disjoint and if the event

$$E_1 \text{ or } E_2 \text{ or } E_3 \text{ or } \ldots = S$$

In other words, every outcome in $S$ is in one and only one of the events $E_1, E_2, \ldots$.

Suppose that $E_1, E_2, \ldots$ form a partition of $S$. Let $A$ be any event in $S$. Then, the **Law of Total Probability** tells us that

$$P(A) = P(A|E_1)P(E_1) + P(A|E_2)P(E_2) + P(A|E_3)P(E_3) + \ldots = \sum_i P(A|E_i)P(E_i)$$

We can verify this formula in the following example. A compulsory statistics module has just been introduced into a university biology class. Last year, there were 240 students in the class, 180 of whom passed the statistics module first time. Letting $A$ = 'randomly selected student passes the module first time', then

$$P(A) = \frac{180}{240} \ (= 0.750)$$

Of the students in the class last year, 100 had taken a statistics course earlier in their time at university. Of these students 89 passed the statistics module first time.

| | Passed module | Failed module | Total |
|---|---|---|---|
| Previous statistics course | 89 | 11 | 100 |
| No previous statistics course | 91 | 49 | 140 |
| Total | 180 | 60 | 240 |

Letting $E_1$ = 'randomly selected student has previously taken a statistics course', then

$$P(A|E_1) = \frac{89}{100} \ (= 0.890)$$

Letting $E_2$ = 'randomly selected student has not previously taken a statistics course', then

$$P(A \mid E_2) = \frac{91}{140} \; (= 0.650)$$

Clearly, $E_1$ and $E_2$ partition the sample space (since $E_2 = E_1'$), and

$$P(E_1) = \frac{100}{240}, \quad P(E_2) = \frac{140}{240}$$

It is easily verified that

$$\begin{aligned} P(A) &= P(A \mid E_1)P(E_1) + P(A \mid E_2)P(E_2) \\ &= \left(\frac{89}{100} \times \frac{100}{240}\right) + \left(\frac{91}{140} \times \frac{140}{240}\right) \\ &= \frac{89 + 91}{240} = \frac{180}{240} \end{aligned}$$

as before.

However, the Law of Total Probability also allows us to project ahead to a different situation. It is hoped in future years to encourage students intending to enter this biology class to take a statistics class in preparation for it. The Department believes that, in future, 80% of students will do this. Projecting ahead, assuming that the conditional probabilities do not change, then

$$\begin{aligned} P(A) &= P(A \mid E_1)P(E_1) + P(A \mid E_2)P(E_2) \\ &= (0.890 \times 0.8) + (0.650 \times 0.2) \\ &= 0.842 \end{aligned}$$

So the Department can look forward to an increased number of students passing the statistics module first time.

Here is another example. Acute pancreatitis is a potentially fatal disorder of the pancreas that occurs in two forms: Type 1 and Type 2. About 60% of all cases of this disorder are Type 1. The events $E_1$ = 'sufferer has Type 1 of this disorder' and $E_2$ = 'sufferer has Type 2 of this disorder' partition the sample space in the random experiment of determining the form of the disorder being experienced by an individual. Note that $P(E_1) = 0.6$ and $P(E_2) = 0.4$.

The causes of the two forms of this disorder, and hence the correct treatment, are quite different, but they can be hard to tell apart just by clinical examination. A biochemical test can be used to help determine which form of the disorder a particular sufferer has. Using this test, about 88% of people suffering from Type 1 are correctly diagnosed as suffering from Type 1, but the remaining 12% are wrongly diagnosed as suffering from Type 2. Similarly, about 84% of people suffering from Type 2 are correctly diagnosed as suffering from Type 2, but the remaining 16% are wrongly diagnosed as suffering from Type 1.

We will now use the Law of Total Probability to determine the probability of the event $A$ = 'individual is diagnosed as suffering from Type 1'. First, it is useful to gather all the probabilities together on the following **tree diagram**.

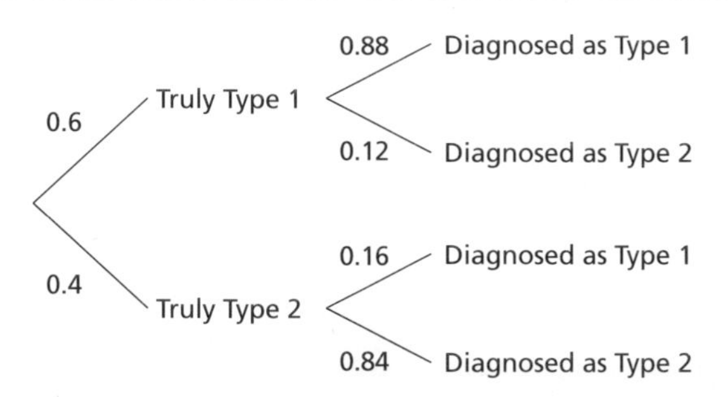


Notice that the *unconditional* probabilities of the events $E_1$ and $E_2$ are shown on the first (furthest left) part of the tree diagram, while the *conditional* probabilities *given* $E_1$ or $E_2$ are shown on the right of the diagram.

The probability associated with any path from left to right through this diagram is found by multiplying together the probabilities on the path. For example, the first path represents the event 'individual is truly Type 1 *and* is diagnosed as Type 1', i.e. the event $E_1$ *and* $A$. Multiplying together the two probabilities on this path,

$$\text{P}(E_1 \textit{ and } A) = \text{P}(A|E_1) \times \text{P}(E_1) = 0.88 \times 0.6$$

Using the Law of Total Probability to evaluate P($A$) is equivalent to adding up the probabilities associated with all the paths through the diagram that lead to event $A$ at the right-hand side. So

$$\begin{aligned}\text{P}(A) &= \text{P}(A|E_1)\text{P}(E_1) + \text{P}(A|E_2)\text{P}(E_2)\\ &= (0.88 \times 0.60) + (0.16 \times 0.40)\\ &= 0.528 + 0.064\\ &= 0.592\end{aligned}$$

In other words, 59.2% of sufferers from acute pancreatitis are diagnosed as suffering from Type 1 form of the disorder. This is just about the correct percentage, but not the correct people. Some people suffering from both forms of the disorder have been misclassified by this test.

So, we might ask a different kind of question. Given that an individual suffering from acute pancreatitis has been diagnosed as having Type 1 form,.what is the conditional probability that this individual really does suffer from the disorder? In other words, what is P($E_1|A$)? Finding a conditional probability of this kind is equivalent to reversing the order of the tree diagram – we now wish to find the conditional probability of one of the events towards the left of the diagram conditional on an event towards the right of the diagram.

To find this out, we need a further result known as **Bayes' Theorem** (after an eighteenth-century clergyman and amateur mathematician, Rev. Thomas Bayes). We can derive this result as follows. Consider two events of non-zero probability, $A$ and $E$. Then, using the definition of conditional probability,

$$\text{P}(A \textit{ and } E) = \text{P}(A|E)\text{P}(E)$$

and

$$\text{P}(E \textit{ and } A) = \text{P}(E|A)\text{P}(A)$$

But, *E and A* is just the same event as *A and E*. This means that

$$P(A|E)P(E) = P(E|A)P(A)$$

Re-arranging this equation, we obtain the result

$$P(E|A) = \frac{P(A|E)P(E)}{P(A)}$$

When $E_1$, $E_2$, ... form a partition of the sample space and $A$ is any other event with non-zero probability then, using the Law of Total Probability,

$$P(E_j|A) = \frac{P(A|E_j)P(E_j)}{P(A)} = \frac{P(A|E_j)P(E_j)}{\sum_i P(A|E_i)P(E_i)} \quad \text{(Bayes' Theorem)}$$

In the example about acute pancreatitis, this means that the conditional probability that an individual truly suffers from Type 1 form of the disorder, when he or she is diagnosed as having Type 1 form, is

$$P(E_1|A) = \frac{P(A|E_1)P(E_1)}{P(A|E_1)P(E_1) + P(A|E_2)P(E_2)}$$

$$= \frac{0.88 \times 0.6}{(0.88 \times 0.6) + (0.16 \times 0.4)} = 0.892$$

So, only 89% of people who are diagnosed as Type 1 actually have Type 1. The remaining 11% actually have Type 2, and there is a real danger that they will be treated inappropriately.

## EXERCISE 1.7B

**1** The proportions of boys and girls who suffer from red–green colour blindness are 0.05 and 0.0025 respectively. 52% of all children in the UK are boys.

**a** Copy the following tree diagram, and complete it by adding the appropriate unconditional and conditional probabilities.

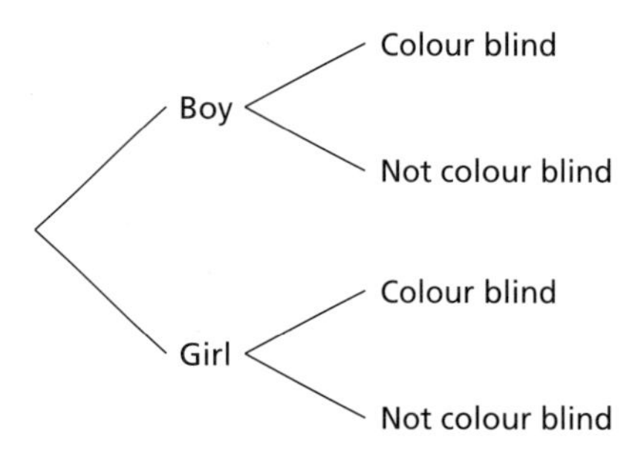


**b** What proportion of children in the UK suffer from red–green colour blindness?

**c** In the UK, what proportion of colour-blind children are girls?

2 Because of a problem that has recently developed in part of the machinery, one of the three production lines in a factory has started to produce a lot of defective items. Just 1% of the items produced on lines A and B are defective, but 4% of the items produced on line C are now defective. If lines A, B and C produce 40%, 30% and 30% (respectively) of the total factory output, find:

**a** the proportion of all the items manufactured in this factory that are defective,

**b** the proportion of all the defective items that are produced on line C.

3 A computer expert offers home support to families who have bought a personal computer. He believes that 10% of his customers have 'excellent' computing skills, that another 65% have 'adequate' skills and that the remaining 25% have 'inadequate' skills. In a given year, he is called to the homes of 20% of those with 'excellent' skills, 30% of those with 'adequate' skills and 50% of those with 'inadequate' skills.

**a** What proportion of his clients call him to their homes in the course of a year?

**b** What proportion of the clients whose homes he visits fall into each of the three categories of computing skills?

4 Ann has reached the final of her Tennis Club championship. In the final, she will play Bernadette, the reigning Club Champion. The final will be won by the first of them to win two sets. Based on her experience of their many previous matches, Ann estimates that she has probability 0.4 of winning the first set. If she wins the first set, she believes she has probability 0.7 of winning any subsequent set. If she loses the first set, on the other hand, she believes she has probability 0.9 of losing any subsequent set.

**a** Find the probability that the final is decided by playing just two sets.

**b** Hence find the probability that the final is decided in three sets.

**c** Given that the final is decided in two sets, find the probability that Ann wins.

5 In a factory that makes fine china ornaments, quality inspectors rate each item as first, second or reject. Two inspectors rate each item, and records suggest that they each rate 60%, 30% and 10% of all ornaments (respectively) into these three categories. The final quality rating of an item is the poorer of the two ratings it receives from the inspectors.

**a** Assuming that the two inspectors give their ratings independently of each other, what proportions of ornaments are finally rated first, second and reject?

**b** Given that an item is finally rated as reject, what is the probability that it was rated reject by both inspectors?

**c** How valid would you expect the assumption of independence to be in this context?

# STATISTICS IN ACTION – DIAGNOSTIC TESTING IN MEDICINE

In an attempt to make possible an early diagnosis of a metabolic disorder in babies, a new screening test has been developed. This test involves determining the amount of a certain steroid that is secreted in the baby's urine. Babies with the metabolic disorder tend to have a raised steroid level. It is intended to use this test as a screening procedure. Any baby who tests positive on it will be referred for further examination. This is just one example of many biochemical tests that are used routinely to aid medical diagnosis. It is extremely important to recognise that almost all such tests are developed using statistical criteria and that there is always a chance that the outcome of the test could be misleading.

Extensive trials suggest that this urine test is positive for 95% of babies who have the disorder (true positives) and that it is negative for 90% of babies who do not have the disorder (true negatives). The remaining babies are wrongly classified. This is a very good test; the misclassification rates can be higher for other routine tests.

The conditional probabilities given above have special names in the medical literature:

P(test is +|baby is truly +) is the **sensitivity** of the test

P(test is –|baby is truly –) is the **specificity** of the test

The performance of the test in practice depends not just on the sensitivity and specificity, but also on the **incidence** of the disorder, $p$ ($0 < p < 1$), which is the proportion of all babies tested who are truly +.

1 Using a tree diagram, find P(baby is truly +|test is +) as a function of $p$.

2 The doctors who have developed the test believe that at first it will be used conservatively, so that $p$ will be about 0.25. Evaluate P(baby is truly +|test is +) in this case. How should a mother react if her baby tests positive under these conditions?

3 This is a rare disorder and the doctors recognise that eventually the test may be used in a wider population of babies, so that $p$ might fall to 0.1. Calculate P(baby is truly +|test is +) under these conditions, and comment.

4 Plot P(baby is truly +|test is +) for values of $p$ between 0.01 and 0.25. Comment on how this probability is affected by $p$.

5 It would be possible to amend the test to change the sensitivity and specificity. Unfortunately, if the sensitivity is increased, the specificity decreases (and vice versa). It would be easily possible to make (a) the sensitivity equal to 0.99 and the specificity equal to 0.80, or (b) the sensitivity equal to 0.90 and the specificity equal to 0.92. Plot P(baby is truly + | test is +) against $p$ for each of these new conditions, and compare your finding with what you discovered in part 4.

6 Finally, show algebraically that one of the conditions (a) and (b) gives a better value of P(baby is truly + | test is +) than the original test for every possible $p$, while the other condition gives a worse value than the original test for every possible $p$. What problem do you foresee with the condition that gives the best result in this comparison?

# CHAPTER 1 SUMMARY

**1** Any process by which information is obtained is called an **experiment**, and the information is called the **outcome** of the experiment. If there are a number of possible outcomes, and it is not possible to tell in advance which of them will occur, the process is called a **random experiment**. Repeats of a random experiment, under identical conditions, are called **trials**.

**2** An exhaustive list of all the possible outcomes of a random experiment is called the **sample space** ($S$). Any collection of outcomes (subset of $S$) is called an **event**. One particular event is the **empty set**, $\{\ \}$, which contains none of the outcomes in $S$.

**3** **(i)** The **complement** of the event $E$, denoted $E'$, is the event that consists of all the outcomes in $S$ that are not in $E$ itself.

**(ii)** The event $E$ *and* $F$ consists of the outcomes that appear in both $E$ and $F$.

**(iii)** The event $E$ *or* $F$ consists of the outcomes that appear either in $E$ or in $F$.

**(iv)** When $E$ *and* $F = \{\ \}$, then the events $E$ and $F$ are said to be **disjoint** or **mutually exclusive**.

**4** **(i)** The **probability** of an event $E$, denoted $\mathrm{P}(E)$, is a measure of how likely the event is to occur. $0 \le \mathrm{P}(E) \le 1$.

**(ii)** $\mathrm{P}(S) = 1$ (certain)
$\mathrm{P}(\{\ \}) = 0$ (impossible)
$\mathrm{P}(E') = 1 - \mathrm{P}(E)$

**(iii)** When $E$ and $F$ are disjoint,

$$\mathrm{P}(E \textit{ and } F) = \mathrm{P}(\{\ \}) = 0$$
$$\mathrm{P}(E \textit{ or } F) = \mathrm{P}(E) + \mathrm{P}(F)$$

**(iv)** In general,

$$\mathrm{P}(E \textit{ or } F) = \mathrm{P}(E) + \mathrm{P}(F) - \mathrm{P}(E \textit{ and } F)$$

**5** When all the outcomes in $S$ are **equally likely**, the probability of any event $E$ is

$$\mathrm{P}(E) = \frac{\text{number of different outcomes in } E}{\text{number of different outcomes in } S}$$

**6** **(i)** $n! = n \times (n-1) \times \ldots \times 1 \quad (n = 1, 2, \ldots)$
$0! = 1$

**(ii)** Any ordered choice of $r$ from $n$ objects is called a **permutation**. The number of different permutations of $r$ from $n$ distinct objects is

$$^{n}P_{r} = \frac{n!}{(n-r)!} \quad (r = 1, 2, \ldots, n)$$

**(iii)** Any unordered choice of $r$ from $n$ objects is called a **combination**. The number of different combinations of $r$ from $n$ distinct objects is

$$\binom{n}{r} = \frac{n!}{r!(n-r)!} \quad (r = 0, 1, 2, \ldots, n)$$

**7** The **conditional probability** of $E$ given $F$ is

$$P(E|F) = \frac{P(E \text{ and } F)}{P(F)} \quad (P(F) > 0)$$

So $$P(E \text{ and } F) = P(E|F)P(F)$$

**8** The events $E$ and $F$ are **independent** if

$$P(E \text{ and } F) = P(E) \times P(F)$$

**9** **(i)** The events $E_1, E_2, \ldots$ **partition** $S$ if the events all have non-zero probability and are all disjoint and if $E_1$ *and* $E_2$ *and* $\ldots = S$.

**(ii)** For any event $A$ in $S$, the **Law of Total Probability** holds:

$$P(A) = \sum_i P(A|E_i)P(E_i)$$

**(iii)** If $P(A) > 0$, **Bayes' Theorem** holds:

$$P(E_j|A) = \frac{P(A|E_j)P(E_j)}{\sum_i P(A|E_i)P(E_i)}$$

# CHAPTER 1 REVIEW EXERCISE

**1** Three standard dice are rolled.

**a** How many (ordered) outcomes are there in the sample space for this experiment?

**b** Find the probability that the scores obtained on the three dice are all the same.

**c** Find the probability that the scores obtained on the three dice are all different.

**d** Hence find the probability that the same score is obtained on two of the dice, but a different score is obtained on the third die.

**2** Three sixth-year pupils are to be chosen to represent the school at a debate being sponsored by the Local Authority. Carl and Mark are among the 20 pupils who have shown an interest in attending this event. In the interests of fairness, the school has decided to choose three of the interested pupils at random to take part in the debate.

**a** Find the probability that Carl will be chosen to take part in the debate.

**b** Find the probability that neither Carl nor Mark is chosen to go to the debate. Hence find the probability that at least one of them is chosen to go.

**c** Now use the general addition rule to find the probability that both Carl and Mark are chosen to take part in the debate.

**3** In a classic experiment, 6952 sweet peas were bred using a 'dihybrid backcross'. Sweet peas of this kind have either purple or red flowers, and their pollen is either long or round. The characteristics of this sample of sweet peas are indicated in the table below.

| | Long pollen | Round pollen | Total |
|---|---|---|---|
| Purple flowers | 4831 | 390 | 5221 |
| Red flowers | 393 | 1338 | 1731 |
| Total | 5224 | 1728 | 6952 |

**a** Assuming that one of the plants in this sample is chosen at random, find the probability that

**(i)** it has long pollen,

**(ii)** it has long pollen, *given* it has purple flowers,

**(iii)** it has long pollen, *given* it has red flowers.

**b** Discuss whether, in sweet peas of this kind, flower colour and pollen shape seem likely to be independent characteristics.

**4** To pass a certain mathematics course, students must perform satisfactorily both in continuously assessed coursework tasks and in a final examination. 12% of students do not complete the coursework tasks satisfactorily, and are therefore disqualified from sitting the final examination. Of those students who sit the examination, 90% perform satisfactorily in it. What percentage of all students who take this course fail it?

**5** A sales representative who deals in heavy machinery calls on either one or two clients each day. On about $\frac{1}{4}$ of days, the representative visits one client, and on the remaining days two clients. The outcome of a visit to a client is a sale, with probability $\frac{1}{3}$, or no sale, with probability $\frac{2}{3}$. The outcomes of visits to different clients are independent. Find the probability that, on a randomly selected day, the representative makes

**a** no sale,

**b** at least one sale.

[Hint: you might find it helpful to construct a tree diagram.]

**6** An experienced trader in computer games is able to classify the second-hand games he is offered for re-sale as either 'very popular', 'popular' or 'unpopular'. Within one month of putting it on display, he expects to sell a 'very popular' game with probability 0.95, a 'popular' game with probability 0.80 and an 'unpopular' game with probability 0.5. Suppose this trader buys a large consignment of second-hand games, of which 20% are 'very popular', 70% are 'popular' and 10% are 'unpopular'.

**a** What proportion of these games will he sell within one month?

**b** Of the games he sells within one month, what proportion will be 'unpopular' games?

# 2 Random Variables



## Review of discrete random variables

A **random variable** is a function which associates a unique real value with each outcome in the sample space of a random experiment.

For example, suppose a random experiment consists of rolling two dice. The sample space consists of 36 ordered pairs of numbers representing the scores on the two dice:

(1, 1) (1, 2) (1, 3) (1, 4) (1, 5) (1, 6)
(2, 1) (2, 2) (2, 3) (2, 4) (2, 5) (2, 6)
(3, 1) (3, 2) (3, 3) (3, 4) (3, 5) (3, 6)
(4, 1) (4, 2) (4, 3) (4, 4) (4, 5) (4, 6)
(5, 1) (5, 2) (5, 3) (5, 4) (5, 5) (5, 6)
(6, 1) (6, 2) (6, 3) (6, 4) (6, 5) (6, 6)

Suppose the random variable $X$ is the number of sixes showing. The only possible values for $X$ are 0, 1 and 2. We say that the **range space** of $X$ is {0, 1, 2}. If the range space of $X$ can be written in the form $\{x_1, x_2, x_3, \ldots\}$ then $X$ is a **discrete random variable**. In this example the range space of $X$ is finite. With each value $x$ of $X$ we may associate the probability of its occurrence $\mathrm{P}(X = x) = \mathrm{p}(x)$.

The set of all pairs $(x, \mathrm{p}(x))$ is called the **probability distribution** of $X$. In this example it is convenient to show the probability distribution of $X$ in a table:

| $x$ | 0 | 1 | 2 |
|---|---|---|---|
| $\mathrm{p}(x)$ | $\frac{25}{36}$ | $\frac{10}{36}$ | $\frac{1}{36}$ |

Notice that $0 \le \mathrm{p}(x) \le 1$ and

$$\sum_{\text{all } x} \mathrm{p}(x) = \mathrm{p}(0) + \mathrm{p}(1) + \mathrm{p}(2) = \frac{25}{36} + \frac{10}{36} + \frac{1}{36} = 1$$

The **cumulative distribution function** of a discrete random variable gives information about $\mathrm{P}(X \le x)$. For this example the values of the cumulative distribution function can be tabulated as follows:

| $x$ | 0 | 1 | 2 |
|---|---|---|---|
| $\mathrm{P}(X \le x)$ | $\frac{25}{36}$ | $\frac{35}{36}$ | 1 |

The cumulative distribution of $X$

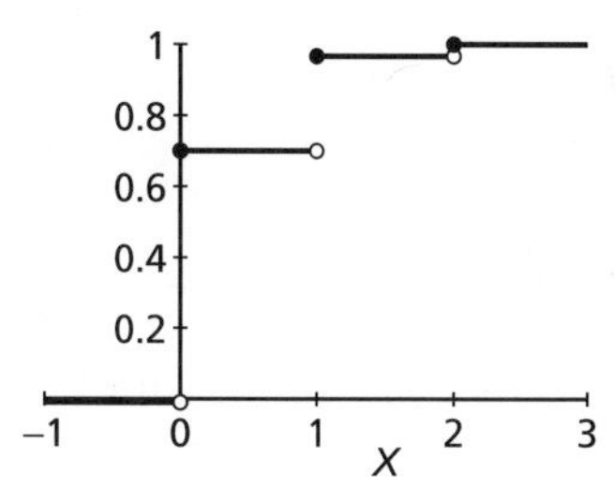


The values of a cumulative distribution function, being probabilities, must lie between 0 and 1 inclusive.

The **mean** $\mu$ or **expected value** $E(X)$ of the discrete random variable $X$ is the average value that would be recorded if the underlying random experiment were carried out a very large number of times and the value of $X$ observed on each occasion.

$$\mu = E(X) = \sum_{\text{all } x} x p(x) \quad (\mu \text{ is a Greek letter pronounced 'mu'})$$

The **variance** $\sigma^2$ or $V(X)$ of a discrete random variable is a measure of how spread out the recorded values of $X$ would be if the underlying random experiment were carried out a very large number of times. Other things being equal, the larger the variance, the more spread out the values of $X$.

$$\sigma^2 = V(X) = E[(X - \mu)^2] = \sum_{\text{all } x} (x - \mu)^2 p(x) \quad (\sigma \text{ is a Greek letter pronounced 'sigma'})$$

Notice that $\sigma^2$ must always be non-negative.

An alternative formula which is easier for calculation is

$$V(X) = E(X^2) - (E(X))^2 \text{ where } E(X^2) = \sum_{\text{all } x} x^2 p(x)$$

The alternative formula can be derived as follows:

$$\begin{aligned} E[(X - \mu)^2] &= \sum (x - \mu)^2 p(x) \\ &= \sum (x^2 - 2\mu x + \mu^2) p(x) \\ &= \sum x^2 p(x) - 2\mu \sum x p(x) + \mu^2 \sum p(x) \\ &= E(X^2) - 2E(X)E(X) + (E(X))^2 1 \quad \text{since } \mu = E(X) \\ &= E(X^2) - (E(X))^2 \end{aligned}$$

The positive square root of the variance, $\sigma$, is known as the **standard deviation**, $sd(X)$.

In our example, where $X$ is the number of sixes showing when two dice are rolled,

$$E(X) = \frac{25}{36} \times 0 + \frac{10}{36} \times 1 + \frac{1}{36} \times 2 = \frac{12}{36} = \frac{1}{3}$$

$$E(X^2) = \frac{25}{36} \times 0^2 + \frac{10}{36} \times 1^2 + \frac{1}{36} \times 2^2 = \frac{14}{36} = \frac{7}{18}$$

$$V(X) = \frac{7}{18} - \left(\frac{1}{3}\right)^2 = \frac{5}{18}$$

and $$sd(X) = \sqrt{\frac{5}{18}} = \frac{\sqrt{10}}{6} \approx 0.527$$

## EXERCISE 2.1A

1 Two dice are rolled and the random variable $T$ is the sum of the two scores showing.
   - **a** Construct a table showing the probability distribution of $T$.
   - **b** Calculate the mean and variance of $T$.

**2** The discrete random variable $Y$ has a uniform distribution given by

$$p(y) = \begin{cases} \frac{1}{n} & \text{for } y = 1, 2, 3, \ldots, n \\ 0 & \text{otherwise} \end{cases}$$

**a** Construct a table showing the probability distribution of $Y$ when $n = 5$.
**b** Calculate the mean and variance of $Y$.
**c** Tabulate the values of the cumulative distribution function and illustrate it with a suitable diagram.

**3** A bag contains nine balls of which five are black and four are white. Three balls are withdrawn without replacement. $X$ is the number of black balls withdrawn.

$$p(x) = \begin{cases} \dfrac{\binom{5}{x}\binom{4}{3-x}}{\binom{9}{3}} & \text{for } x = 0, 1, 2, 3 \\ 0 & \text{otherwise} \end{cases}$$

**a** Show the probability distribution of $X$ in a table.
**b** Calculate the mean and variance of $X$.

**4** A game consists of rolling a fair die and if a six appears the die is rolled a second time. A player's score is a random variable $Z$ representing the total number of spots which appeared. The range space of $Z$ is $\{1, 2, 3, 4, 5, 7, 8, 9, 10, 11, 12\}$. Write down the probability distribution of $Z$ and hence calculate the mean and variance of $Z$. [Hint: you might find it helpful to draw a tree diagram for this example.]

**5** The discrete random variable $X$ has the range space $\{1, 2, 3, 4\}$ and a cumulative distribution function whose values can be tabulated as follows:

| $x$ | 1 | 2 | 3 | 4 |
|---|---|---|---|---|
| $P(X \le x)$ | 0.1 | 0.3 | 0.6 | 1.0 |

**a** Tabulate the probability distribution of $X$.
**b** Calculate the mean, variance and standard deviation of $X$.

**6 a** Use the fact $\Sigma_{\text{all } x} p(x) = 1$ to find the value of $k$ so that

$$p(x) = \begin{cases} k\binom{5}{x} & \text{for } x = 0, 1, 2, 3, 4, 5 \\ 0 & \text{otherwise} \end{cases}$$

could serve as the probability distribution of a discrete random variable $X$.
**b** Calculate the mean and variance of $X$.
**c** Tabulate the values of the cumulative distribution function of $X$.

## EXERCISE 2.1B

**1** A gymnast is learning a new routine. Her probability of failure on the first attempt is 0.8, on the second attempt the probability of failure decreases to 0.6 and on her third attempt the probability of failure is 0.4. The random variable $Z$ represents the number of successfully completed routines during her first three attempts.

**a** Construct a tree diagram to show the possible outcomes of her three practice attempts.

**b** Construct a table of the probability distribution of $Z$.

**c** Calculate the mean, variance and standard deviation of $Z$.

**2** The discrete random variable $X$ has range space $\{1, 2, 3, 4, 5\}$ and its cumulative distribution function takes the value

$$P(X \leq x) = \frac{x^2 + 5x}{50}, \quad x = 1, 2, 3, 4, 5$$

**a** Tabulate the probability distribution of $X$.

**b** Calculate $E(X)$ and $V(X)$.

**3** When tossed, a particular coin lands Heads with probability $p$. The discrete random variable $X$ represents the number of Heads obtained when this coin is tossed twice.

**a** Tabulate the probability distribution of $X$.

**b** Show that $E(X) = 2p$ and $V(X) = 2p(1 - p)$.

**c** Evaluate the mean and variance of $X$ for a coin which is twice as likely to land Tails than Heads.

**4** A random variable $Y$ has the following probability distribution

| $y$ | $-2$ | $0$ | $2$ |
|---|---|---|---|
| $P(Y = y)$ | $k$ | $1 - 2k$ | $k$ |

**a** Show that $V(Y) = 8k$.

**b** Calculate the variance and illustrate the probability distribution using a bar diagram when **(i)** $k = \frac{1}{8}$ **(ii)** $k = \frac{1}{3}$ **(iii)** $k = \frac{1}{2}$.

**5 a** Show that the mean and variance of the discrete uniform random variable $X$ with

$$P(X = x) = \begin{cases} \dfrac{1}{n} & x = 1, 2, 3, \ldots, n \\ 0 & \text{otherwise} \end{cases}$$

are $E(X) = \frac{1}{2}(n + 1)$ and $V(X) = \frac{1}{12}(n^2 - 1)$ respectively. You may use the following results:

$$\sum_{x=1}^{n} x = \frac{1}{2}n(n + 1), \quad \sum_{x=1}^{n} x^2 = \frac{1}{6}n(n + 1)(2n + 1)$$

**b** Verify these formulae for the mean and variance of $X$ when $X$ represents the score on a fair die.

**c** A special die is made with its faces labelled $-2, -1, 0, 1, 2, 3$, and each score is equally likely. What would be the mean and variance of the scores produced by this die?

**d** If the range space for a discrete uniform random variable $X$ is $k + 1, \ldots, k + n$ where $k$ is any integer, suggest formulae for calculating the mean and variance of $X$.

**e** In theory the random digits 0, 1, 2, …, 9 produced by a certain computer program follow a uniform distribution. Calculate the expected value and variance of the digits it produces.



## Laws of expectation and variance

Suppose a random experiment consists of rolling a die. Let the random variable $X$ be the score on the die. In this case the sample space of the random experiment and the range space of the random variable are both {1, 2, 3, 4, 5, 6}.

$$E(X) = 1\times\frac{1}{6} + 2\times\frac{1}{6} + 3\times\frac{1}{6} + 4\times\frac{1}{6} + 5\times\frac{1}{6} + 6\times\frac{1}{6} = \frac{7}{2}$$

$$V(X) = 1^2\times\frac{1}{6} + 2^2\times\frac{1}{6} + 3^2\times\frac{1}{6} + 4^2\times\frac{1}{6} + 5^2\times\frac{1}{6} + 6^2\times\frac{1}{6} - \left(\frac{7}{2}\right)^2 = \frac{35}{12}$$

It is possible to create new random variables which are related to $X$ in various ways.

### Adding a constant: W = X + b

Suppose $W = X + 3$.
The probability distribution for $W$ is

| $w$ | 4 | 5 | 6 | 7 | 8 | 9 |
|---|---|---|---|---|---|---|
| p($w$) | $\frac{1}{6}$ | $\frac{1}{6}$ | $\frac{1}{6}$ | $\frac{1}{6}$ | $\frac{1}{6}$ | $\frac{1}{6}$ |

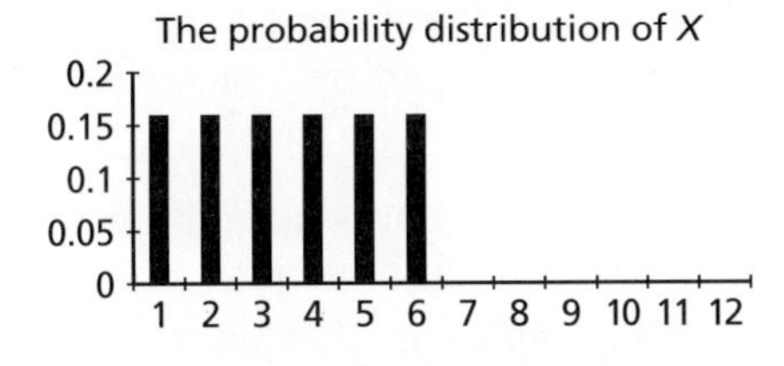


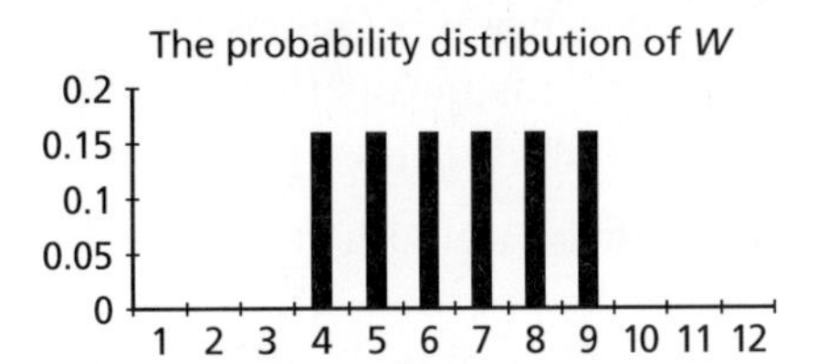


$$\begin{aligned} E(X + 3) &= E(W) \\ &= 4\times\frac{1}{6} + 5\times\frac{1}{6} + 6\times\frac{1}{6} + 7\times\frac{1}{6} + 8\times\frac{1}{6} + 9\times\frac{1}{6} \\ &= \frac{13}{2} \\ &= E(X) + 3 \end{aligned}$$

$$\begin{aligned} V(X + 3) &= V(W) \\ &= 4^2\times\frac{1}{6} + 5^2\times\frac{1}{6} + 6^2\times\frac{1}{6} + 7^2\times\frac{1}{6} + 8^2\times\frac{1}{6} + 9^2\times\frac{1}{6} - \left(\frac{13}{2}\right)^2 \\ &= \frac{35}{12} \\ &= V(X) \end{aligned}$$

It can be shown that in general $E(X + b) = E(X) + b$ and that $V(X + b) = V(X)$.

Adding a constant to a random variable has the effect of shifting the location of the expected value or mean but has no effect on the variance or spread. This can be seen in the diagrams above.

### *Multiplying by a constant: W = aX*

Suppose $W = 2X$.
The probability distribution for $W$ is

| $w$ | 2 | 4 | 6 | 8 | 10 | 12 |
|---|---|---|---|---|---|---|
| $p(w)$ | $\frac{1}{6}$ | $\frac{1}{6}$ | $\frac{1}{6}$ | $\frac{1}{6}$ | $\frac{1}{6}$ | $\frac{1}{6}$ |

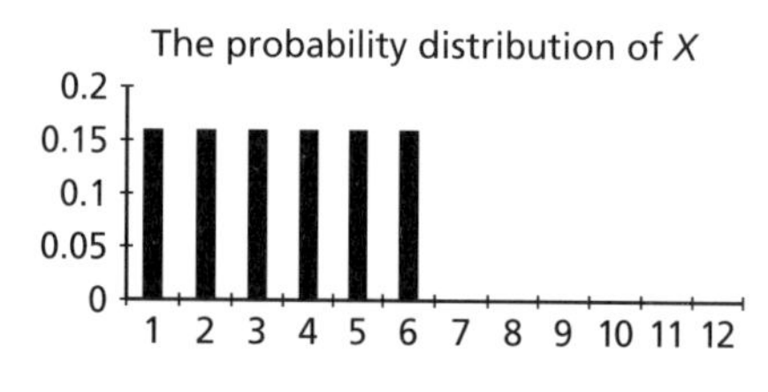


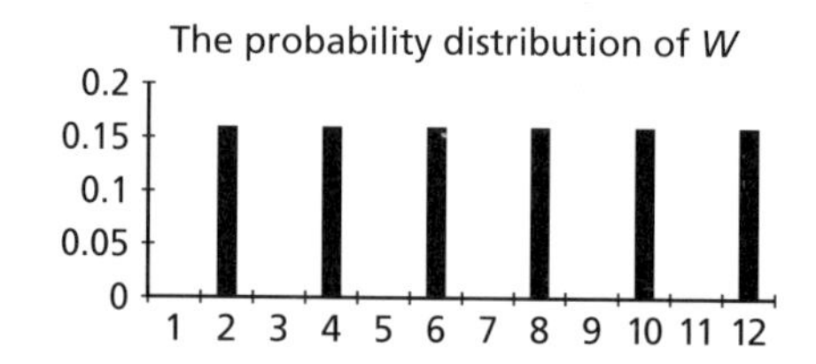


$$\begin{aligned} E(2X) &= E(W) \\ &= 2\times\frac{1}{6} + 4\times\frac{1}{6} + 6\times\frac{1}{6} + 8\times\frac{1}{6} + 10\times\frac{1}{6} + 12\times\frac{1}{6} \\ &= 7 \\ &= 2E(X) \end{aligned}$$

$$\begin{aligned} V(2X) &= V(W) \\ &= 2^2\times\frac{1}{6} + 4^2\times\frac{1}{6} + 6^2\times\frac{1}{6} + 8^2\times\frac{1}{6} + 10^2\times\frac{1}{6} + 12^2\times\frac{1}{6} - 7^2 \\ &= \frac{35}{3} \\ &= 4V(X) \end{aligned}$$

It can be shown that in general $E(aX) = aE(X)$ and that $V(aX) = a^2V(X)$.

Multiplying a random variable by a constant has the effect of a change of scale. The expected value or mean is changed by a factor of $a$, while the variance is changed by a factor of $a^2$. This means that the standard deviation is changed by a factor of $a$. This can be seen in the diagrams above.

Suppose now that a random experiment consists of spinning a square spinner at the same time as a die is rolled. The square spinner, with edges numbered 1, 2, 3 and 4 (respectively), comes to rest with an edge resting on a flat table top. As before, $X$ is the score on the die. Let the random variable $Y$ be the score on the spinner. The range space of $Y$ is {1, 2, 3, 4} and each value of $Y$ is equally likely.

$$E(Y) = 1\times\frac{1}{4} + 2\times\frac{1}{4} + 3\times\frac{1}{4} + 4\times\frac{1}{4} = \frac{5}{2}$$

and $$V(Y) = 1^2\times\frac{1}{4} + 2^2\times\frac{1}{4} + 3^2\times\frac{1}{4} + 4^2\times\frac{1}{4} - \left(\frac{5}{2}\right)^2 = \frac{5}{4}$$

### *Adding random variables: X + Y*

If all 24 equally-likely outcomes of the random experiment are listed as pairs $(x, y)$ then we can easily write beside each pair the value of $X + Y$.

| | | | |
|---|---|---|---|
| (1, 1)**2** | (1, 2)**3** | (1, 3)**4** | (1, 4)**5** |
| (2, 1)**3** | (2, 2)**4** | (2, 3)**5** | (2, 4)**6** |
| (3, 1)**4** | (3, 2)**5** | (3, 3)**6** | (3, 4)**7** |
| (4, 1)**5** | (4, 2)**6** | (4, 3)**7** | (4, 4)**8** |
| (5, 1)**6** | (5, 2)**7** | (5, 3)**8** | (5, 4)**9** |
| (6, 1)**7** | (6, 2)**8** | (6, 3)**9** | (6, 4)**10** |

Thus, the probability distribution of $X + Y$ is

| $x + y$ | 2 | 3 | 4 | 5 | 6 | 7 | 8 | 9 | 10 |
|---|---|---|---|---|---|---|---|---|---|
| $p(x + y)$ | $\frac{1}{24}$ | $\frac{2}{24}$ | $\frac{3}{24}$ | $\frac{4}{24}$ | $\frac{4}{24}$ | $\frac{4}{24}$ | $\frac{3}{24}$ | $\frac{2}{24}$ | $\frac{1}{24}$ |

$$\begin{aligned} E(X+Y) &= 2\times\frac{1}{24} + 3\times\frac{2}{24} + 4\times\frac{3}{24} + 5\times\frac{4}{24} + 6\times\frac{4}{24} + 7\times\frac{4}{24} + 8\times\frac{3}{24} + 9\times\frac{2}{24} + 10\times\frac{1}{24} \\ &= \frac{144}{24} = 6 \\ &= E(X) + E(Y) \end{aligned}$$

$$\begin{aligned} V(X+Y) &= 2^2\times\frac{1}{24} + 3^2\times\frac{2}{24} + 4^2\times\frac{3}{24} + 5^2\times\frac{4}{24} + 6^2\times\frac{4}{24} + 7^2\times\frac{4}{24} + 8^2\times\frac{3}{24} + 9^2\times\frac{2}{24} \\ &\quad + 10^2\times\frac{1}{24} - 6^2 \\ &= \frac{964}{24} - 36 = \frac{25}{6} \\ &= V(X) + V(Y) \end{aligned}$$

This result for the variances is only true if the random variables $X$ and $Y$ are independent.

The discrete random variables, $X$ and $Y$, are said to be **independent** if

$$P(X = x \textit{ and } Y = y) = P(X = x) \times P(Y = y)$$

for every possible choice of $x$ and $y$. Note the analogy with the definition of independence for two events. Here the events '$X = x$' and '$Y = y$' are required to be independent for every possible choice of $x$ and $y$. This is sometimes called the **product model** for $X$ and $Y$.

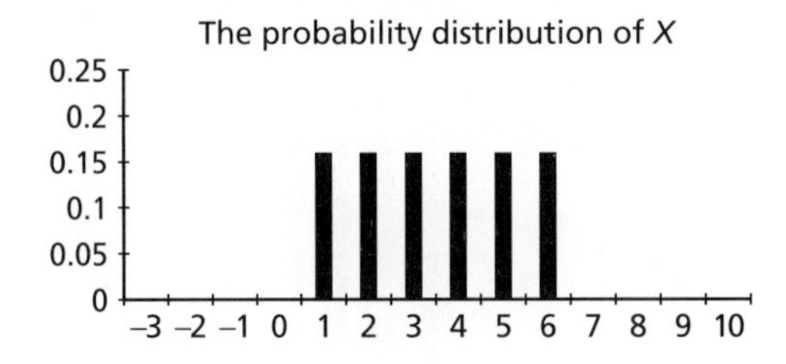


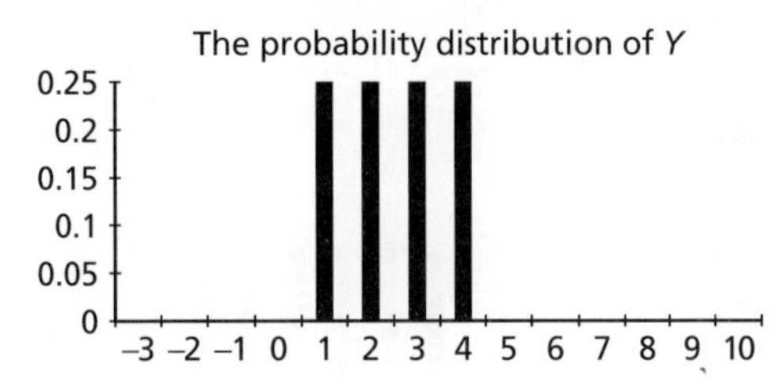


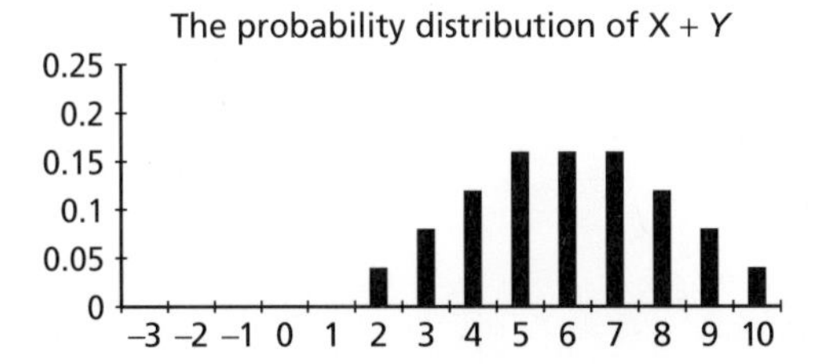


It can be shown that in general if $X$ and $Y$ are any two random variables then

$$E(X + Y) = E(X) + E(Y)$$

and

$$V(X + Y) = V(X) + V(Y)$$

provided $X$ and $Y$ are independent.

## *Subtracting random variables: X − Y*

If all the possible outcomes of the random experiment are listed as pairs $(x, y)$ then we can easily write beside each pair the value of $X - Y$.

| | | | |
|---|---|---|---|
| (1, 1)**0** | (1, 2)**−1** | (1, 3)**−2** | (1, 4)**−3** |
| (2, 1)**1** | (2, 2)**0** | (2, 3)**−1** | (2, 4)**−2** |
| (3, 1)**2** | (3, 2)**1** | (3, 3)**0** | (3, 4)**−1** |
| (4, 1)**3** | (4, 2)**2** | (4, 3)**1** | (4, 4)**0** |
| (5, 1)**4** | (5, 2)**3** | (5, 3)**2** | (5, 4)**1** |
| (6, 1)**5** | (6, 2)**4** | (6, 3)**3** | (6, 4)**2** |

Thus, the probability distribution of $X - Y$ is

| $x - y$ | −3 | −2 | −1 | 0 | 1 | 2 | 3 | 4 | 5 |
|---|---|---|---|---|---|---|---|---|---|
| $p(x - y)$ | $\frac{1}{24}$ | $\frac{2}{24}$ | $\frac{3}{24}$ | $\frac{4}{24}$ | $\frac{4}{24}$ | $\frac{4}{24}$ | $\frac{3}{24}$ | $\frac{2}{24}$ | $\frac{1}{24}$ |

$$\begin{aligned}E(X - Y) &= -3\times\frac{1}{24} - 2\times\frac{2}{24} - 1\times\frac{3}{24} + 0\times\frac{4}{24} + 1\times\frac{4}{24} + 2\times\frac{4}{24} + 3\times\frac{3}{24} + 4\times\frac{2}{24} + 5\times\frac{1}{24}\\ &= \frac{24}{24} = 1\\ &= E(X) - E(Y)\end{aligned}$$

$$\begin{aligned}V(X - Y) &= (-3)^2\times\frac{1}{24} + (-2)^2\times\frac{2}{24} + (-1)^2\times\frac{3}{24} + 0^2\times\frac{4}{24} + 1^2\times\frac{4}{24} + 2^2\times\frac{4}{24} + 3^2\times\frac{3}{24}\\ &\quad + 4^2\times\frac{2}{24} + 5^2\times\frac{1}{24} - 1^2\\ &= \frac{124}{24} - 1 = \frac{25}{6}\\ &= V(X) + V(Y)\end{aligned}$$

This result for the variances is only true if the random variables $X$ and $Y$ are independent.

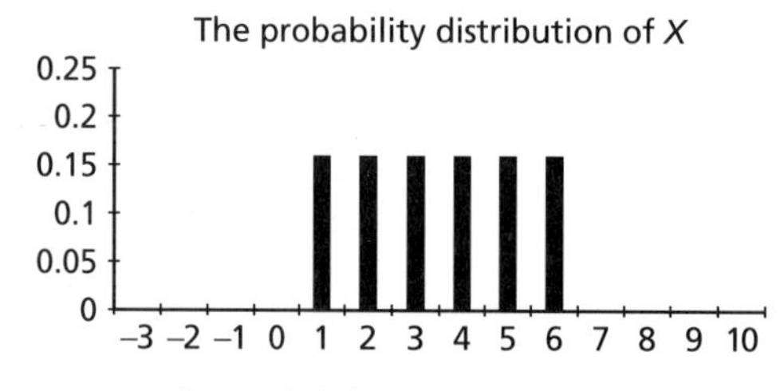


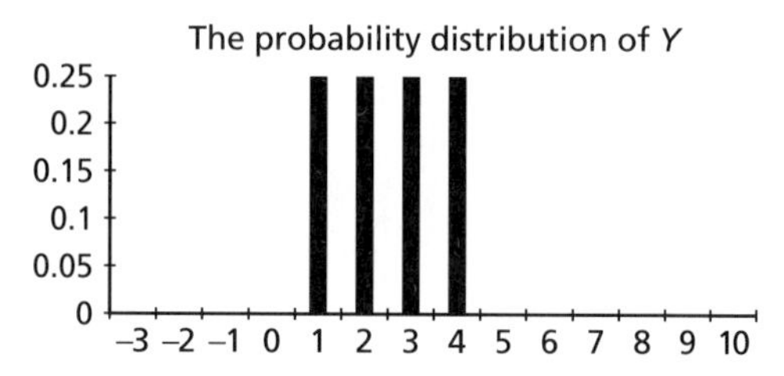


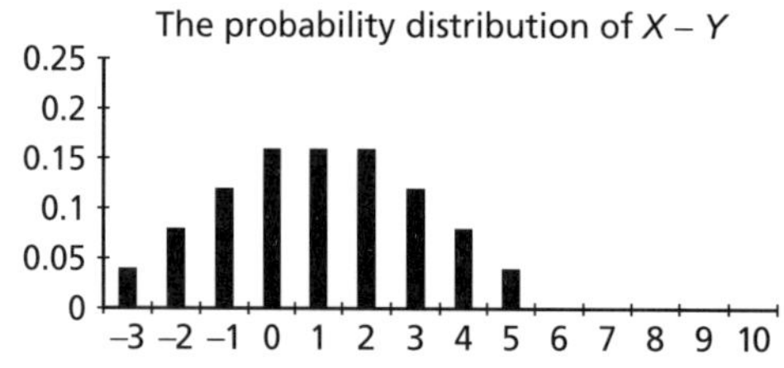


It can be shown that in general if $X$ and $Y$ are any two random variables then

$$E(X - Y) = E(X) - E(Y)$$

and

$$V(X - Y) = V(X) + V(Y)$$

provided $X$ and $Y$ are independent.

***Summary***

$E(aX + b) = aE(X) + b$.

$V(aX + b) = a^2V(X)$.

$E(X \pm Y) = E(X) \pm E(Y)$.

$V(X \pm Y) = V(X) + V(Y)$, provided $X$ and $Y$ are independent.

## EXERCISE 2.2

**1** The random variable $Y$ represents the score which results when a four-sided spinner comes to rest. The range space of $Y$ is $\{1, 2, 3, 4\}$ and each value of $Y$ is equally likely.

**a** Tabulate the probability distribution for $W$ where

**(i)** $W = Y + 2$ **(ii)** $W = 3Y$ **(iii)** $W = 3Y + 2$

**b** For each case in part **a**, calculate the mean and variance of $W$ from its probability distribution.

**c** Hence show that

**(i)** $E(Y + 2) = E(Y) + 2$, $V(Y + 2) = V(Y)$ **(ii)** $E(3Y) = 3E(Y)$, $V(3Y) = 9V(Y)$ **(iii)** $E(3Y + 2) = 3E(Y) + 2$, $V(3Y + 2) = 9V(Y)$

**2** A random variable $X$ has mean $\mu = 5$ and variance $\sigma^2 = 4$.
Calculate the mean and variance of the random variable $W$ when

**a** $W = X - 3$ **b** $W = 4X$ **c** $W = 2X - 5$

**3** Two fair coins are tossed. For the first coin, the random variable $X$ takes the value +1 when it shows a Head and 0 when it shows a Tail. The random variable $Y$ takes the value +1 when the second coin shows a Head and 0 when it shows a Tail.

**a** Construct the probability distribution for **(i)** $X + Y$ **(ii)** $X - Y$ **(iii)** $3X - 2Y$.

**b** From the probability distributions, calculate the mean and variance for each of the three combinations of $X$ and $Y$ in part **a**.

**c** Hence show that

**(i)** $E(X + Y) = E(X) + E(Y)$ and $V(X + Y) = V(X) + V(Y)$

**(ii)** $E(X - Y) = E(X) - E(Y)$ and $V(X - Y) = V(X) + V(Y)$

**(iii)** $E(3X - 2Y) = 3E(X) - 2E(Y)$ and $V(3X - 2Y) = 9V(X) + 4V(Y)$

**4** Two fair dice are rolled. The random variable $X$ represents the score on the first die and $Y$ represents the score on the second die. Draw bar diagrams to illustrate the probability distributions of **a** $X$, **b** $Y$, **c** $X + Y$.
Below each diagram write the appropriate mean and variance.

**5** Two independent random variables, $X$ and $Y$, have means 20 and 15 and standard deviations 4 and 3 respectively. Find the mean, variance and standard deviation of

**a** $3X - 2$ **b** $X - Y$

**6** The random variable $X$ has mean 15 and standard deviation 4 while the independent random variable $Y$ has mean 10 and standard deviation 3.
Calculate the mean and standard deviation of each of the following:

**a** $X + Y$ **b** $X - Y$ **c** $4X - 3Y$ **d** $2X + Y - 5$

7 Two fair dice are rolled. The random variable $X$ represents the score on the first die and $Y$ represents the score on the second die.

**a** Construct the probability distribution for $XY$.

**b** Calculate $E(XY)$.

c Show that $E(XY) = E(X)E(Y)$.

(Note that this result is only true when $X$ and $Y$ are independent.)



## *The Binomial distribution*

This discrete probability distribution is one of the most important distributions in the whole of statistics. The conditions that give rise to a Binomial distribution are:

(i) there is a *fixed* number of $n$ trials;
(ii) only *two* outcomes, 'success' and failure', are possible at each trial;
(iii) the trials are *independent*;
(iv) there is a *constant probability* $p$ of success;
(v) the random variable, $X$, is the *total number of successes* in $n$ trials.

The random variable $X$ is said to follow a Binomial distribution, i.e. $X \sim \text{Bin}(n, p)$. With the above conditions, it can be shown that $X$ has probability distribution

$$P(X = x) = \begin{cases} \binom{n}{x} p^x q^{n-x} & x = 0, 1, 2, \ldots, n \\ 0 & \text{otherwise} \end{cases} \qquad \text{where } 0 < p < 1 \text{ and } q = 1 - p$$

For example, suppose a random experiment consists of tossing a fair coin three times and the random variable $X$ represents the number of Heads which appear. Each of the above conditions applies:

(i) there is a fixed number of trials, $n = 3$;
(ii) there are only two outcomes possible at each trial, 'Heads' and 'Tails';
(iii) trials are independent because the result when a coin is tossed is not affected by the results of previous tosses;
(iv) the probability of getting a 'Head' remains constant, $p = 0.5$;
(v) the random variable $X$ represents the total number of 'Heads' in three tosses.

It is appropriate to use the model $X \sim \text{Bin}\left(3, \frac{1}{2}\right)$ and we can calculate probabilities using the formula:

$$P(X = 0) = \binom{3}{0}\left(\frac{1}{2}\right)^0\left(\frac{1}{2}\right)^3 = 1 \times 1 \times \frac{1}{8} = \frac{1}{8}$$

$$P(X = 1) = \binom{3}{1}\left(\frac{1}{2}\right)^1\left(\frac{1}{2}\right)^2 = 3 \times \frac{1}{2} \times \frac{1}{4} = \frac{3}{8}$$

$$P(X = 2) = \binom{3}{2}\left(\frac{1}{2}\right)^2\left(\frac{1}{2}\right)^1 = 3 \times \frac{1}{4} \times \frac{1}{2} = \frac{3}{8}$$

$$P(X = 3) = \binom{3}{3}\left(\frac{1}{2}\right)^3\left(\frac{1}{2}\right)^0 = 1 \times \frac{1}{8} \times 1 = \frac{1}{8}$$

| $x$ | 0 | 1 | 2 | 3 |
|---|---|---|---|---|
| $P(X = x)$ | $\frac{1}{8}$ | $\frac{3}{8}$ | $\frac{3}{8}$ | $\frac{1}{8}$ |

$$E(X) = 0\times\frac{1}{8} + 1\times\frac{3}{8} + 2\times\frac{3}{8} + 3\times\frac{1}{8} = \frac{3}{2}$$

$$E(X^2) = 0^2\times\frac{1}{8} + 1^2\times\frac{3}{8} + 2^2\times\frac{3}{8} + 3^2\times\frac{1}{8} = 3$$

$$V(X) = 3 - \left(\frac{3}{2}\right)^2 = \frac{3}{4}$$

In general, it can be shown that if $X \sim \text{Bin}(n, p)$ then

$$E(X) = np \text{ and } V(X) = np(1 - p) = npq.$$

These formulae provide a quick way to calculate the mean and variance of a Binomial random variable. Check that they give the same answer in this example.

Table 1 of cumulative Binomial probabilities (Appendix 1, page 135) can be used for certain values of $n$ and $p$. For example, suppose $R \sim \text{Bin}(6, 0.4)$. The required section of Table 1 is:

| $p$ | 0.05 | 0.10 | 0.15 | 0.20 | 0.25 | 0.30 | 0.35 | 0.40 | 0.45 | 0.50 |
|---|---|---|---|---|---|---|---|---|---|---|
| $n = 6$ $x = 0$ | | | | | | | | 0.0467 | | |
| 1 | | | | | | | | 0.2333 | | |
| 2 | | | | | | | | 0.5443 | | |
| 3 | | | | | | | | 0.8208 | | |
| 4 | | | | | | | | 0.9590 | | |
| 5 | | | | | | | | 0.9959 | | |

These are cumulative probabilities. Individual probabilities are calculated as follows:

$$P(X = 0) = 0.0467$$
$$P(X = 1) = 0.2333 - 0.0467 = 0.1866$$
$$P(X = 2) = 0.5443 - 0.2333 = 0.3110$$

and so on until

$$P(X = 5) = 0.9959 - 0.9590 = 0.0369$$
$$P(X = 6) = 1 - 0.9959 = 0.0041$$

Notice that Table 1 lists cumulative probabilities only for certain values of $p$ up to $p = 0.5$. However, it is possible to extend the use of Table 1 for values beyond 0.5. For example, suppose $Y \sim \text{Bin}(6, 0.8)$ where $Y$ can be thought of as the number of 'successes' in six trials. Let $X$ be the number of 'failures' in the same six trials. Then, $X = 6 - Y$ but $X \sim \text{Bin}(6, 0.2)$ and we can look up the cumulative probabilities of $X$ as before.

| $p$ | 0.05 | 0.10 | 0.15 | 0.20 | 0.25 | 0.30 | 0.35 | 0.40 | 0.45 | 0.50 |
|---|---|---|---|---|---|---|---|---|---|---|
| $n = 6$ $x = 0$ | | | | 0.2621 | | | | | | |
| 1 | | | | 0.6554 | | | | | | |
| 2 | | | | 0.9011 | | | | | | |
| 3 | | | | 0.9830 | | | | | | |
| 4 | | | | 0.9984 | | | | | | |
| 5 | | | | 0.9999 | | | | | | |

Using these values, as before, we can list the probability distribution of $X$:

$p(0) = 0.2621 \quad p(1) = 0.3933 \quad p(2) = 0.2457 \quad p(3) = 0.0819$

$p(4) = 0.0154 \quad p(5) = 0.0015 \quad p(6) = 0.0001$

The probability distribution of $Y$ is now found by noting that $P(Y = y) = P(X = 6 - y)$.

We can obtain the probabilities for $Y$ by reversing the list of probabilities for $X$:

$p(0) = 0.0001 \quad p(1) = 0.0015 \quad p(2) = 0.0154 \quad p(3) = 0.0819$

$p(4) = 0.2457 \quad p(5) = 0.3933 \quad p(6) = 0.2621$

An efficient method of calculating Binomial probabilities is to make use of a recurrence formula. For example, suppose $X \sim \text{Bin}(5, 0.4)$ and we wish to list the probability distribution of $X$.

$$P(X = 0) = \binom{5}{0}0.4^0 0.6^5 = 0.077\,76$$

$$P(X = 1) = \binom{5}{1}0.4^1 0.6^4 = P(X = 0) \times \frac{5}{1} \times \frac{0.4}{0.6} = 0.2592$$

$$P(X = 2) = \binom{5}{2}0.4^2 0.6^3 = P(X = 1) \times \frac{4}{2} \times \frac{0.4}{0.6} = 0.3456$$

$$P(X = 3) = \binom{5}{3}0.4^3 0.6^2 = P(X = 2) \times \frac{3}{3} \times \frac{0.4}{0.6} = 0.2304$$

$$P(X = 4) = \binom{5}{4}0.4^4 0.6^1 = P(X = 3) \times \frac{2}{4} \times \frac{0.4}{0.6} = 0.0768$$

$$P(X = 5) = \binom{5}{5}0.4^5 0.6^0 = P(X = 4) \times \frac{1}{5} \times \frac{0.4}{0.6} = 0.010\,24$$

**Note**
As a useful check, make sure all the probabilities add up to 1.

This recurrence method of calculating Binomial probabilities can be expressed as follows:

If $X \sim \text{Bin}(n, p)$ then

$$P(X = 0) = q^n$$

and

$$P(X = x) = \frac{n + 1 - x}{x} \times \frac{p}{q} \times P(X = x - 1) \text{ for } x = 1, 2, \ldots, n$$

## EXERCISE 2.3A

**1** $X \sim \text{Bin}(9, 0.3)$. Calculate **a** $P(X = 2)$ **b** $P(X = 5)$ **c** $P(X = 8)$

**2** Using Table 1 (page 135), list the first four values of the probability distributions of:
**a** $W \sim \text{Bin}(10, 0.35)$ **b** $X \sim \text{Bin}(12, 0.5)$ **c** $Y \sim \text{Bin}(16, 0.40)$

**3** $X \sim \text{Bin}(5, 0.3)$ and $Y \sim \text{Bin}(5, 0.7)$. Calculate the probability distributions for $X$ and $Y$. What do you notice?

**4** Extend the use of Table 1 and list the first four values of the probability distributions of:
**a** $R \sim \text{Bin}(8, 0.6)$ **b** $T \sim \text{Bin}(12, 0.55)$ **c** $V \sim \text{Bin}(10, 0.9)$

5 What is the probability that exactly five Heads appear when a fair coin is tossed
   a 12 times   b 15 times? (Use Table 1 where possible.)

6 For each of the following random variables, state, with reasons, whether the Binomial distribution would provide a satisfactory model. If it would, state the values of $n$ and $p$.
   a The number of Heads obtained in four successive tosses of a coin.
   b The number of Heads obtained when four coins are tossed simultaneously.
   c The number of boys in the families of British monarchs.
   d The number of Queens in a hand of four cards dealt from a standard pack of cards.
   e The number of children in a class of 25 whose birthday falls on a Friday this year.
   f The number of throws required when a die is rolled until the first six appears.
   g The number of women in random groups of six successive people chosen as people leave a cinema. You may assume that the cinema audience was large and that there were approximately equal numbers of men and women in the audience.

7 It is known that 5% of electronic components of a certain kind are defective. A random sample of ten components is taken from a large stockpile.
   a What is the probability that there are at most two defective components in the sample? What assumption have you made?
   b Calculate the mean and variance of the number of defective components in such samples.

8 It is believed that in a certain population 40% of people have blood type A. A random sample of 20 people are selected from this population.
   a Calculate the probability that five or fewer people in this sample have blood type A. What assumption have you made?
   b Calculate the mean and standard deviation of the number of people in such samples with blood type A.

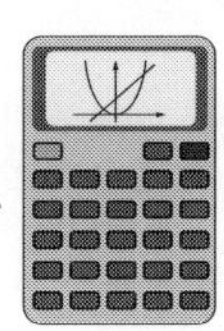

Some calculators will give you Binomial probabilities.
Look for menu items such as **binompdf(** and **binomcdf(**.
Find out what your calculator can do.

## EXERCISE 2.3B

1 Use the recurrence formula to list the first four values of the probability distribution of:
   a $X \sim \text{Bin}(5, 0.4)$   b $Y \sim \text{Bin}(7, 0.6)$   c $Z \sim \text{Bin}(9, 0.55)$

2 Extend the use of Table 1 and list the first four values of the probability distribution of:
   a $K \sim \text{Bin}(8, 0.65)$   b $L \sim \text{Bin}(6, 0.7)$   c $M \sim \text{Bin}(10, 0.75)$

3 A group of eight patients who have all been diagnosed as suffering from a particular disease are recruited to a clinical trial of a new drug. They are allocated at random with equal probability to receive either the new drug or a placebo (an inactive substance which looks the same as the drug). Calculate the probability that
   a exactly four patients receive the new drug,
   b all eight patients receive the placebo.

**4** Individual cards from a set of ten picture cards are distributed at random in boxes of cereal. A collector needs just one particular card to complete the set. What is the probability that she will complete her set if she buys:

**a** 1 box **b** 5 boxes **c** 20 boxes

**5** A multiple-choice test consists of ten questions and there are four responses to each question (of which only one is correct). A student answers each question by guessing.

**a** What is the probability that he gets exactly five correct answers?

**b** Calculate the mean and variance of the number of correct answers he achieves.

**6** Airline overbooking is common practice. Past records show that 20% of people making a reservation fail to show up. An airline operates a busy commuter service between two cities using small aircraft that have 16 seats. They regularly accept 18 reservations for each flight. Let $X$ be the random variable representing the number of passengers who show up.

**a** Assuming passengers act independently, state the distribution of $X$ and its range space.

**b** Calculate the probability that, on any given flight, at least one passenger holding a reservation will not have a seat.

**c** Determine the mean and standard deviation of $X$.

**7** Illustrate the distribution of $X \sim \text{Bin}(10, p)$ by means of a bar chart when:

**a** $p = 0.25$ **b** $p = 0.5$ **c** $p = 0.75$

Compare these bar charts and describe their shape.

**8** Draw a series of bar charts to illustrate $X \sim \text{Bin}(n, 0.25)$ when

**a** $n = 4$ **b** $n = 10$ **c** $n = 20$

Describe any pattern you see in these diagrams.

## *The Poisson distribution*

The Poisson distribution is an important model for discrete random variables whose range space is the set of whole numbers. The conditions that give rise to a Poisson distribution are:

(i) *isolated events* are occurring in *continuous* time or space;

(ii) the numbers of events that occur in non-overlapping segments of time or space are *independent*;

(iii) events occur *singly* rather than in groups;

(iv) events are occurring at a *constant average rate* per unit time or space throughout the whole period or region of interest;

(v) the random variable, $X$, is the number of events occurring per segment.

The random variable $X$ is said to follow a Poisson distribution, i.e. $X \sim \text{Poi}(\mu)$. With the above conditions, it can be shown that $X$ has probability distribution

**Reminder**
The number $e \approx 2.718$

$$P(X = x) = \frac{e^{-\mu}\mu^x}{x!} \quad \text{for } x = 0, 1, 2, 3, \ldots$$

For example, the number of telephone calls received at a particular switchboard during business hours is thought to be a Poisson random variable with an average of eight calls every 5 minutes. The above criteria are satisfied:

(i) the arrival of a telephone call at the switchboard is an isolated event in continuous time;

(ii) the numbers of telephone calls received in different 5 minute non-overlapping segments of time may be assumed to be independent (note that this may not be the case if the switchboard receives telephone calls in response to an appeal to a television audience);

(iii) if telephone calls are initiated by individuals acting independently then we may assume that these events are occurring singly rather than in groups;

(iv) we assume that the telephone calls arrive at a constant average rate of eight calls every 5 minutes during business hours;

(v) the random variable is the number of calls received at the switchboard in a 5 minute segment.

When $X \sim \text{Poi}(8)$ we may calculate probabilities of individual values of $X$ using the formula above. For example,

$$\text{P}(X=0) = \frac{e^{-8} \times 8^0}{0!} = e^{-8} = 0.0003, \quad \text{P}(X=6) = \frac{e^{-8} \times 8^6}{6!} = 0.1221$$

Alternatively, for certain values of $\mu$ we can use Table 2 of cumulative Poisson probabilities (Appendix 1, page 138). From this table we have that for $X \sim \text{Poi}(8)$

$$\text{P}(X=0) = 0.0003, \quad \text{P}(X=1) = 0.0030 - 0.0003 = 0.0027,$$
$$\text{P}(X=2) = 0.0138 - 0.0030 = 0.0108, \ldots$$

Note that for $X \sim \text{Poi}(8)$ in Table 2, cumulative probabilities are listed correct to four decimal places so

$$\text{P}(X \leq 21) = \text{P}(X \leq 22) = \text{P}(X \leq 23) = 1.0000$$

It can be shown that for a Poisson distribution

$$\text{E}(X) = \text{V}(X) = \mu$$

In this example the mean and variance are both eight calls per 5 minute segment.

An efficient method for calculating Poisson probabilities is to make use of a recurrence formula.

$$\text{P}(X=x) = \frac{e^{-\mu}\mu^x}{x!} = \frac{e^{-\mu}\mu^{x-1}}{(x-1)!} \times \frac{\mu}{x}$$

so we can write

$$\text{P}(X=x) = \text{P}(X=x-1) \times \frac{\mu}{x}, \; x = 1, 2, \ldots$$

with $\text{P}(X=0) = e^{-\mu}$.

For example, the first five values of $X \sim \text{Poi}(1.2)$ are:

$$\text{P}(X=0) = \frac{e^{-1.2}(1.2)^0}{0!} = e^{-1.2} = 0.3012$$

$$\text{P}(X=1) = 0.3012 \times \frac{1.2}{1} = 0.3614$$

$$\text{P}(X=2) = 0.3614 \times \frac{1.2}{2} = 0.2169$$

$$\text{P}(X=3) = 0.2169 \times \frac{1.2}{3} = 0.0867$$

$$\text{P}(X=4) = 0.0867 \times \frac{1.2}{4} = 0.0260$$

## EXERCISE 2.4A

1 $X \sim \text{Poi}(0.6)$. Calculate:
   a $P(X = 0)$ b $P(X = 2)$ c $P(X = 4)$
   Give your answers correct to four decimal places.

2 Using Table 2 (page 138), list the first five values of the probability distribution of:
   a $R \sim \text{Poi}(3.0)$ b $T \sim \text{Poi}(5.5)$ c $V \sim \text{Poi}(9.0)$.

3 Use the recurrence formula to calculate the first six values of $X \sim \text{Poi}(12)$.
   Give your answers correct to four decimal places.

4 A biologist believes that the distribution of a certain plant in a grassland habitat follows a Poisson distribution with mean seven plants per square metre. Using Table 2, determine the probability that in a randomly chosen square metre of grassland there are:
   a fewer than seven plants,
   b exactly seven plants,
   c more than seven plants.

5 On average, a local branch of the British Red Cross receives 2.3 requests for a wheelchair each weekday. Calculate the probability that on a certain weekday they receive:
   a no requests,
   b at most two requests,
   c more than two requests.

6 From past experience it is known that the *Daily Gazette* newspaper has an average of two misprints per page. Use Table 2 to determine the probability that:
   a the front page has no misprints,
   b the back page has two misprints,
   c the second page has fewer than four misprints.

7 The random variable $X$ represents the number of goals scored per match in a particular soccer league. From past experience it is believed that $X \sim \text{Poi}(2.5)$.
   a What is the probability that a randomly chosen match ends in a no-score draw?
   b Assuming the goals scored in different matches are independent, how many no-score draws would you expect on a day when there were 15 league matches played?

8 Do you think the following random variables should be modelled by the Poisson, Binomial, or some other probability distribution? Give reasons for your choice.
   a The number of southbound vehicles passing a bridge on the motorway in a 5 minute interval.
   b The number of girls in a family of four children.
   c The number of schools in a $500\,\text{m} \times 500\,\text{m}$ area of a large city.
   d The number of times a fair die has to be rolled until a 6 appears.
   e The number of major repairs required in a house in any year in the UK.
   f The number of defective electronic components in a box of 100.

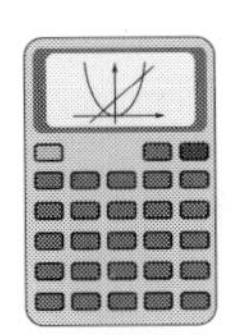

Some calculators will give you Poisson probabilities.
Look for menu items such as **poissonpdf(** and **poissoncdf(**.
Find out what your calculator can do.



## EXERCISE 2.4B

**1** Use the recurrence formula to draw up a table for the first six values of the probability distribution of $X \sim \text{Poi}(2.8)$. Give your answers correct to four decimal places.

**2** A certain type of photocopier has, on average, 1.2 breakdowns per month. Calculate the probability that for a particular month this photocopier will function

**a** without a breakdown,

**b** with exactly two breakdowns.

**3** The number of radioactive particles detected by a Geiger counter is a Poisson random variable with a mean of 5 per minute.

**a** What is the probability that $k$ particles arrive in any 1 minute interval?

**b** Using tables calculate the probability that the number of particles which arrive at the counter in a 1 minute interval is

**(i)** 1, **(ii)** 2, **(iii)** 5, **(iv)** more than 5.

**4** A certain type of fabric has, on average, five flaws per 100 square metres. Assuming the number of flaws follows a Poisson distribution, what is the probability that a 150 square metre piece of fabric will have at least six flaws?

**5** The random variable $X$ follows a Poisson distribution and $P(X = 0) = 0.110\,803\,2$. Calculate:

**a** $E(X)$ and $V(X)$ **b** $P(X \le 3)$

**6** The independent random variables $X$ and $Y$ have Poisson distributions with means $\mu_1$ and $\mu_2$ respectively. The random variable $W = X + Y$.

**a** Show that:

**(i)** $P(W = 0) = e^{-(\mu_1 + \mu_2)}$

**(ii)** $P(W = 1) = (\mu_1 + \mu_2)e^{-(\mu_1 + \mu_2)}$

**b** Determine expressions for:

**(i)** $P(W = 2)$

**(ii)** $P(W = 3)$

**c** Deduce the distribution of $W$.

**7** Draw bar charts to illustrate $X \sim \text{Poi}(\mu)$ when

**a** $\mu = 1$ **b** $\mu = 2$ **c** $\mu = 4$

In each diagram, use a vertical probability scale ranging from 0 to 0.4 and draw bars for the first 12 values of $X$. Describe how the shape of the distribution changes as $\mu$ increases.

## *Approximating the Binomial with the Poisson*

Sometimes, in situations where the Binomial distribution is the appropriate model to use, an approximation based on the Poisson distribution is used instead to make the calculations easier.

*Example* From past records it is known that 4% of the items made by a factory are defective. The random variable $X$ represents the number of defective items in a random sample of 100 items taken from the factory's output. Calculate the probability that such a sample contains two defects:

**a** exactly using the Binomial distribution,

**b** using the Poisson approximation to the Binomial.

*Solution*

**a** $X \sim \text{Bin}(100, 0.04)$

$$P(X = 2) = \binom{100}{2} \times 0.04^2 \times 0.96^{98} = 4950 \times 0.0016 \times 0.0183 = 0.1450$$

**b** $\mu = np = 100 \times 0.04 = 4$

$X \sim \text{Poi}(4)$ approximately

$$P(X = 2) = \frac{e^{-4} \times 4^2}{2!} = 8e^{-4} = 0.1465$$

Note that the two answers are quite close.

*Rule of thumb*

If $n$ is large (say $n \geq 20$) and $p$ is small (say $p \leq 0.05$) then Poi($np$) provides a good approximation to Bin($n$, $p$).

The approximation is very good when $n \geq 100$ and $np < 10$.

### EXERCISE 2.5B

**1** $X \sim \text{Bin}(50, 0.01)$. Calculate the first three values of $X$:

**a** using the Binomial formula, **b** using the Poisson approximation.

**2** A random experiment consists of tossing five fair coins simultaneously. The random variable $X$ is the number of times five Heads appear in 100 repetitions of the experiment. Calculate $P(X \leq 1)$ using

**a** the Binomial formula, **b** the Poisson approximation.

**3** The random variable $X$ represents the number of people in a group of 100 whose birthday is on 1 April. Assuming that births are evenly distributed throughout the year and that there are 365 days in a year, calculate the probability that exactly one person in the group has his/her birthday on 1 April using

**a** the Binomial formula, **b** the Poisson approximation.

**4** The probability that an item manufactured on a certain production line is defective is 0.02. Assuming independence between items manufactured on this production line, calculate approximately the probability that, of the next 500 items produced, more than 10 are defective. Comment on the assumption of independence.

**5** A particular type of colour blindness is thought to affect five out of every 100 males. Calculate approximately the probability that there are more than five out of 80 first-year boys who have this type of colour blindness. What assumptions have you made?

**6** A car manufacturer reckons that there is a 1 in 10 000 chance that the petrol tank on one of its models will explode when the car is involved in a certain type of collision. Suppose that this model of car were to be involved in 300 such collisions. Calculate approximately the probability that there will be at least one accident where the tank explodes.

## *Review of continuous random variables*

Some random experiments result in a real number value being recorded rather than an integer value. For example, measuring the heights of a sample of 14-year-old boys can be thought of as a random experiment which results in real number values. Height is an example of a **continuous random variable**. As a teenage boy or girl grows, their height can take any value in some range of values. There are no gaps in this measurement scale even though for convenience we may record heights to the nearest millimetre. Other measurements such as weight, temperature and time also require continuous random variables.

For a discrete random variable we were able to list all the values it took and to assign probabilities to these values. However, since a continuous random variable can take any value over a certain range of values, it is not possible to list them all. Consequently, determining probabilities for a continuous random variable requires a different approach. Probability statements for a continuous random variable $X$ refer to probability over a range of values of $X$. For example:

| | Discrete random variable $D$ | Continuous random variable $X$ |
|---|---|---|
| Range space | $\{1, 2, 3, 4, 5, 6\}$ | $\{x: 100 \leq x \leq 200, x \in R\}$ |
| Probability statements | $P(D = 3) = \frac{1}{6}$<br>or $P(2 \leq D \leq 4) = \frac{1}{2}$ | $P(X < 150) = 0.48$<br>or $P(120 < X < 125) = 0.11$ |

In this example, $P(X = 122) = 0$ because it is not possible to assign probabilities to specific values of $X$. In general, for a continuous random variable, $P(X = x) = 0$. Consequently, $P(X < x) = P(X \leq x)$ etc.

For continuous random variables, probabilities can be determined by first making an assumption about the nature of the population concerned. We assume that the population can be described by a curve having a particular shape.

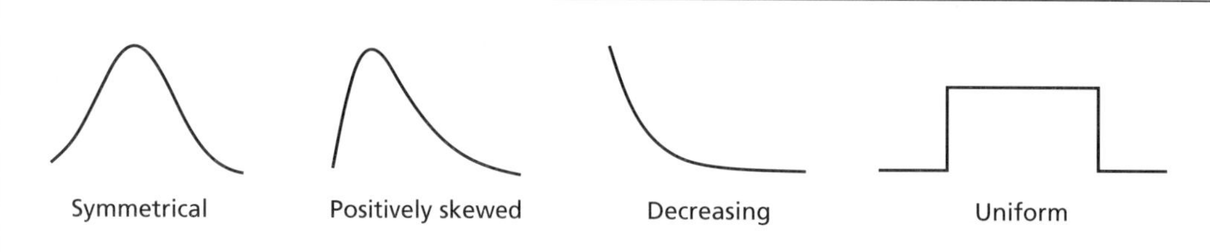


Some common shapes are shown in the diagram. These curves show how the total probability is spread across the range of values that the continuous random variable can take. The curve is a graph showing the random variable's **probability density function** (pdf for short).

The particular shape chosen to model the population may result from theoretical considerations or perhaps from looking at the shape of a histogram of the data in a sample drawn from the population.

For example, suppose a bus company operates a service on a busy city route where the buses are every 10 minutes. The advertising for this service says that a timetable is not necessary because those wishing to use the service will never have to wait longer than 10 minutes.

Let the random variable $T$ represent the time in minutes a person has to wait until the next bus arrives. The range space of $T$ is $\{t: 0 < t < 10, t \in R\}$, i.e. $T$ can take any real number between 0 and 10.

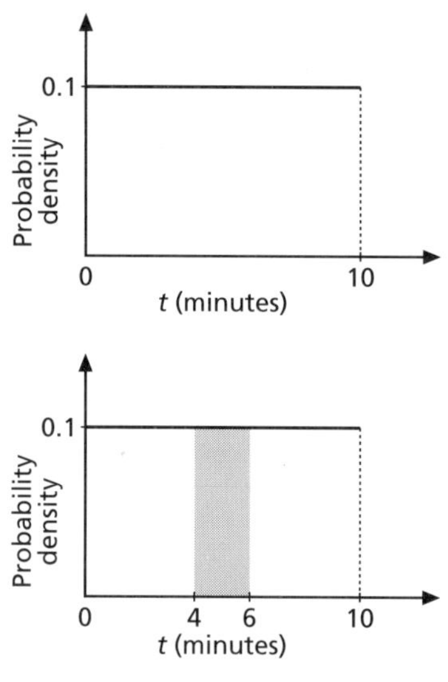


The uniform distribution shown here provides a suitable model for the population of waiting times. The probability density function can be written

$$f(t) = \begin{cases} 0.1 & 0 < t < 10 \\ 0 & \text{otherwise} \end{cases}$$

Probabilities are assigned to a range of values of the random variable $T$ by calculating the area under the curve for the interval of interest.

For example, the probability that someone has to wait between 4 and 6 minutes is $P(4 < T < 6) = 0.2$. Note that the total area under the probability density function is 1.

The **cumulative distribution function** (cdf) is

$$F(t) = P(T \leq t) = \begin{cases} 0 & t \leq 0 \\ 0.1t & 0 < t < 10 \\ 1 & t \geq 10 \end{cases}$$

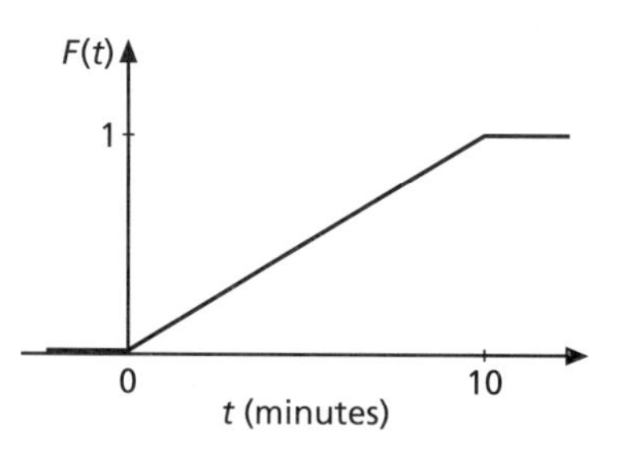


Calculating the areas representing probabilities for a uniform distribution is very straightforward. However, consider the following example.

The proportion of new businesses that fail each year in a local enterprise zone is thought to be a random variable $X$ with the following probability density function

$$f(x) = \begin{cases} 12x(1-x)^2 & 0 < x < 1 \\ 0 & \text{otherwise} \end{cases}$$

The graph of the pdf shows that, unlike the uniform distribution, probability is not spread evenly over the values that the random variable takes. Probabilities are represented by areas under the curve and these are calculated by integration. The total area under the curve is 1.

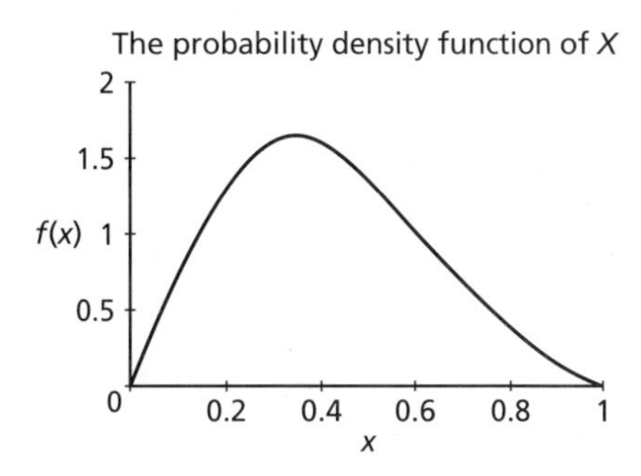


*Example* Using the above model for the proportion of new businesses that fail,

**a** determine the cumulative distribution function,

**b** calculate the probability that less than
(i) 20% (ii) 40% of new businesses fail,

**c** calculate the probability that between 20% and 40% of new businesses fail,

**d** calculate the mean and standard deviation for the proportion of new business failures.

*Solution*

**a** $X$ represents the proportion of new businesses that fail

$$P(X \le x) = \int_0^x f(u)du = \int_0^x 12u(1-u)^2du = 12\int_0^x (u - 2u^2 + u^3)du = 6x^2 - 8x^3 + 3x^4$$

($u$ is a dummy variable, since $x$ has been used as the upper limit of integration)

so that

$$F(x) = \begin{cases} 0 & x \le 0 \\ 6x^2 - 8x^3 + 3x^4 & 0 < x < 1 \\ 1 & x \ge 1 \end{cases}$$

**b** **(i)** $F(0.2) = 6(0.2)^2 - 8(0.2)^3 + 3(0.2)^4 = 0.1808$.
**(ii)** Similarly $F(0.4) = 0.5248$.

**c** $P(0.2 < X < 0.4) = F(0.4) - F(0.2) = 0.5248 - 0.1808 = 0.3440$

**d** $\mu = E(X) = \int_0^1 xf(x)dx = \int_0^1 12x^2(1-x)^2dx = 0.4$

$$E(X^2) = \int_0^1 x^2f(x)dx = \int_0^1 12x^3(1-x)^2dx = 0.2$$

$$V(X) = 0.2 - 0.4^2 = 0.04 \qquad sd(X) = \sqrt{V(X)} = \sqrt{0.04} = 0.2$$

***Summary***

If a continuous random variable $X$ has probability density function

$$\begin{cases} f(x) & a < x < b \\ 0 & \text{otherwise} \end{cases}$$

then the cumulative distribution function of $X$ which gives $P(X \le x)$ is

$$F(x) = \int_a^x f(u)\,du$$

Conversely,

$$f(x) = \frac{d}{dx}F(x)$$

Note that $f(x) \ge 0$ for all values of $x$ and $\int_a^b f(x)\,dx = 1$.

$$P(x_1 < X < x_2) = F(x_2) - F(x_1) \quad \text{or} \quad P(x_1 < X < x_2) = \int_{x_1}^{x_2} f(x)\,dx$$

The mean $\mu$ or expected value of a continuous random variable $X$ is given by

$$\mu = E(X) = \int_a^b x f(x)\,dx$$

and the variance $\sigma^2$ or $V(X)$ is given by

$$V(X) = E(X^2) - (E(X))^2 \quad \text{where } E(X^2) = \int_a^b x^2 f(x)\,dx$$


## EXERCISE 2.6A

**1** A continuous random variable $X$ has the probability density function:

$$f(x) = \begin{cases} 3x^2 & 0 \le x \le 1 \\ 0 & \text{otherwise} \end{cases}$$

**a** Determine the cumulative distribution function.

**b** Hence or otherwise, calculate

**(i)** $P(X \le 0.6)$ **(ii)** $P(X \le 0.3)$ **(iii)** $P(0.3 \le X \le 0.6)$

**c** Determine the mean and variance of $X$.

**2** The cumulative distribution function for a continuous random variable $X$ is:

$$F(x) = \begin{cases} 0 & x < 0 \\ \frac{1}{4}x^2 & 0 \le x \le 2 \\ 1 & x > 1 \end{cases}$$

**a** Find the probability density function $f(x)$ for $X$.

**b** Draw sketches of $F(x)$ and $f(x)$.

**c** Calculate the mean and variance of $X$.

**3** A continuous random variable $X$ has the probability density function:

$$f(x) = \begin{cases} \dfrac{1}{k} & 0 \le x \le 5 \\ 0 & \text{otherwise} \end{cases}$$

Find **a** the value of $k$ **b** the cdf $F(x)$

**c** **(i)** $P(X = 1)$ **(ii)** $P(X \le 1)$ **(iii)** $P(X < 1)$ **(iv)** $P(X > 1)$

**4** The rate at which a tank is filled is a random variable $R$ which is uniformly distributed between 30 and 40 litres per minute.

**a** Calculate $P(33 < R < 35)$.

**b** Write down $P(33 \le R \le 35)$.

**c** Calculate the mean and variance of $R$.

The probability density function of $R$
0.15
0.1
0.05
0
10 20 30 40 50
Litres per minute

**5** The time $X$ seconds between consecutive vehicles passing a fixed point on a motorway is thought to be a random variable with probability density function

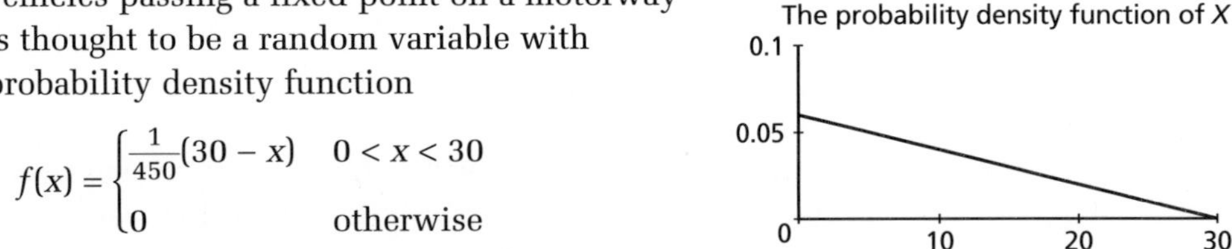


$$f(x) = \begin{cases} \dfrac{1}{450}(30 - x) & 0 < x < 30 \\ 0 & \text{otherwise} \end{cases}$$

**a** Find and sketch $F(x)$, the cdf of $X$.

**b** Calculate $P(10 < X < 20)$.

**c** Determine the mean and variance of $X$.

**d** Calculate $E(Y)$ and $V(Y)$ when **(i)** $Y = 2X$ **(ii)** $Y = 0.5X + 10$.

**6** The percentage of alcohol in a certain chemical compound is $100X\%$ where $X$ is a random variable with probability density function $f(x) = kx(1 - x)$, $0 < x < 1$.

**a** Find the value of $k$ that makes this a valid pdf. [Hint: the total area must equal 1.]

**b** Calculate the mean and standard deviation of $X$.

**c** Determine the cumulative distribution function of $X$.

**d** Calculate the probability that the percentage of alcohol is:
**(i)** less than 10%, **(ii)** more than 60%, **(iii)** between 25% and 50%.

**7** The cumulative distribution function of the random variable $W$ is given by:

$$F(w) = \begin{cases} 1 - \dfrac{16}{w^2} & w > 4 \\ 0 & \text{otherwise} \end{cases}$$

**a** Calculate $P(W \le 6)$ and $P(W > 8)$.

**b** Determine the pdf of $W$ and use it to recalculate the probabilities in part **a**.

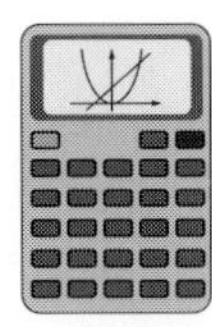

Some calculators can draw the graph of a probability density function and calculate the value of a definite integral. These features could be used when working with continuous random variables. Find out what your calculator can do.

## EXERCISE 2.6B

**1** A component in an automatic washing machine works for $X$ hours without repair or replacement. It is thought that the random variable $X$ has pdf

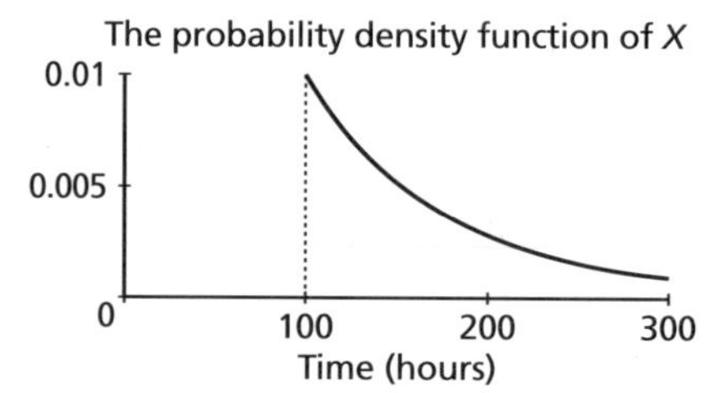


$$f(x) = \begin{cases} \dfrac{100}{x^2} & x > 100 \\ 0 & \text{otherwise} \end{cases}$$

**a** Determine the cumulative distribution function.

**b** Find the median of $X$. [Hint: find $m$ such that $F(m) = 0.5$.]

**c** Calculate the interquartile range (IQR).

**2** A continuous random variable $X$ has probability density function:

$$f(x) = \begin{cases} 2(1 - x) & 0 < x < 1 \\ 0 & \text{otherwise} \end{cases}$$

**a** Show that $E(X^r) = \dfrac{2}{(r+1)(r+2)}$, for $r = 1, 2, 3, \ldots$

**b** Hence or otherwise, determine $E[(2X + 1)^2]$.

**3** A computer has been programmed to draw isosceles triangles in which the sides of equal length are 10 cm long and the angle between them is a random variable, $X$ radians, with pdf:

$$f(x) = \begin{cases} kx(\pi - x) & 0 < x < \pi \\ 0 & \text{otherwise} \end{cases}$$

**a** Determine the value of $k$.

**b** Find the cumulative distribution function of $X$.

**c** Calculate the probability that the third side of a triangle drawn by the computer has length greater than 10 cm.

**4** The temperature (in °C) of the water in a large tank is thought to be modelled by the random variable $T$ with probability density function

$$f(t) = \begin{cases} kt(t - 30) & 0 < t < 30 \\ 0 & \text{otherwise} \end{cases}$$

**a** Determine the value of $k$.

**b** Sketch the pdf and write down $E(T)$.

**c** Calculate the variance of $T$.

**d** Find $E(T)$ and $V(T)$ in °F.

> *Hint*
>
> Recall that $F = \frac{9}{5}C + 32$ where the temperature is $C$ in Celsius and $F$ in Fahrenheit.

**5** A continuous uniform random variable $X$ has the probability density function:

$$f(x) = \begin{cases} \dfrac{1}{b - a} & a < x < b \\ 0 & \text{otherwise} \end{cases}$$

Determine the mean and variance of $X$.

You may use the result $b^3 - a^3 = (b - a)(b^2 + ab + a^2)$.

### *Optional questions*

*For question 6 you need to know how to integrate the exponential function:* Mathematics 1 (AH).
*For question 7 you need to know how to integrate by parts:* Mathematics 2 (AH).



**6** A manufacturer of screens for computer displays has determined that the lifetime, in years, for the picture tube is a random variable $T$ with pdf:

$$f(t) = \begin{cases} 0.1e^{-0.1t} & t > 0 \\ 0 & \text{otherwise} \end{cases}$$

**a** Sketch the pdf.
**b** What proportion of tubes: **(i)** fail within the two-year warranty period **(ii)** last longer than fifteen years?

**7** The continuous random variable $X$ has an exponential distribution with probability density

$$f(x) = \begin{cases} ke^{-kx} & x > 0 \\ 0 & \text{otherwise} \end{cases}$$

Determine the mean and standard deviation of $X$.

## *The Normal distribution*

Many continuous random variables have been found to follow Normal distributions, for example:

the heights of adult Scotsmen
the errors associated with certain astronomical observations
the weights of 200 g packs of a particular brand of bacon

A continuous random variable $X$ is said to have a Normal distribution if it has the probability density function

$$f(x) = \frac{1}{\sqrt{2\pi\sigma^2}}\exp\left(-\frac{(x-\mu)^2}{2\sigma^2}\right), \qquad -\infty < x < \infty$$

We write $X \sim \mathrm{N}(\mu, \sigma^2)$.

The Normal distribution has two parameters

$\mu = \mathrm{E}(X)$ and $\sigma^2 = \mathrm{V}(X)$

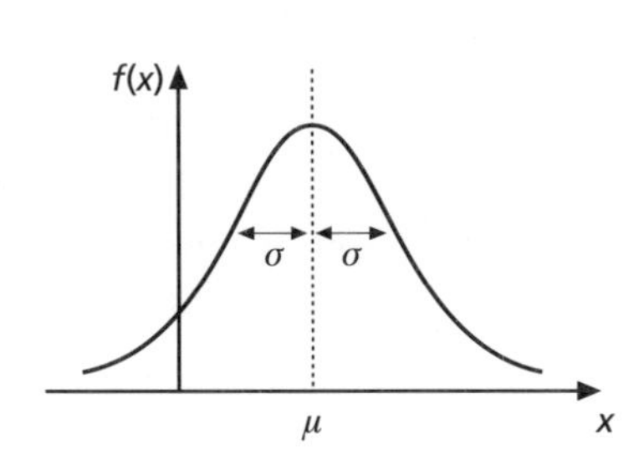


Notice the symmetric 'bell shape' where the probability density is greatest close to the mean.

Unlike the Binomial and Poisson distributions, the mean of a Normal distribution can be changed without altering the variance, and vice versa. In this way we could have many Normal distributions, all with the same shape (variance), which differ only in location (mean), as shown.

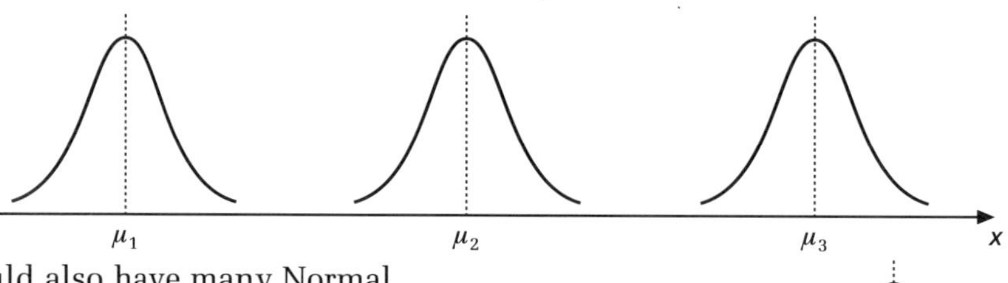


We could also have many Normal distributions all with the same mean, but some more spread out than others, i.e. having differing variances. Clearly, an infinite number of Normal distributions are possible.

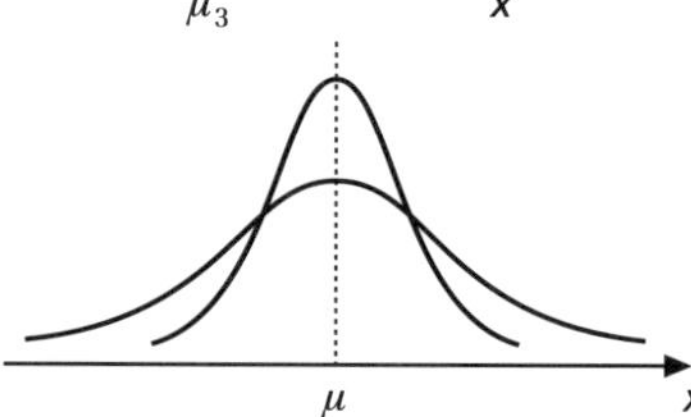


We have seen that, for continuous random variables, probabilities are determined by calculating areas under the probability density curve. Unfortunately, it is not possible to integrate the Normal's pdf by applying simple rules like those we use when working with functions which involve rational powers of $x$. The required areas have to be calculated using numerical integration (see Numerical Analysis 1 (AH).) This has been done for the **Standard Normal** distribution $Z \sim N(0, 1)$ which has a mean of zero and a standard deviation of 1. The cdf of the Standard Normal distribution, often denoted $\Phi(z)$, appears in Table 3(Appendix 1, page 139).

Any random variable $X \sim N(\mu, \sigma^2)$ can be transformed to the Standard Normal random variable $Z$:

$$Z = \frac{X - \mu}{\sigma} \sim N(0, 1)$$

so that

$$P(X < x) = P\left(Z < \frac{x - \mu}{\sigma}\right) = \Phi\left(\frac{x - \mu}{\sigma}\right)$$

Thus, the required probabilities for any Normal distribution can be determined by first using this transformation and then looking up Table 3. The following examples show how this is done and how Table 3 can be extended by making use of the Normal curve's symmetry.

*Example 1* $Z \sim N(0, 1)$. Calculate:

**a** $\Phi(1.27)$ **b** $P(Z > 1.56)$ **c** $\Phi(-0.83)$ **d** $P(Z > -2.24)$

*Solution*

**a** $\Phi(z)$ is the notation used for the cumulative Standard Normal distribution function, i.e. $\Phi(1.27)$ means $P(Z \le 1.27)$ where $Z \sim N(0, 1)$. Table 3 gives values of $\Phi(z)$ for $0 \le z \le 3.69$ in increments of 0.01. In row 1.2 column .07 we find 0.8980, so $\Phi(1.27) = P(Z \le 1.27) = 0.8980$.

**b** $$P(Z \le 1.56) + P(Z > 1.56) = 1$$
$$\text{So } P(Z > 1.56) = 1 - P(Z \le 1.56) = 1 - \Phi(1.56) = 1 - 0.9406 = 0.0594$$

**c** Because the Standard Normal distribution has bilateral symmetry about $z = 0$, Table 3 can be used to determine $\Phi(z)$ for values of $z < 0$. By symmetry,

$$P(Z \le -0.83) = P(Z \ge 0.83) \quad \text{and} \quad P(Z \ge 0.83) = 1 - P(Z \le 0.83)$$

so $\Phi(-0.83) = 1 - \Phi(0.83) = 1 - 0.7967 = 0.2033$

**d** By symmetry,
$P(Z > -2.24) = P(Z < 2.24)$
$= \Phi(2.24) = 0.9875$

*Reminder*
$P(Z \le z) = P(Z < z)$ and $P(Z = z) = 0$

*Example 2* Suppose that $X \sim N(8, 4)$. Find:
**a** $P(X < 10)$ **b** $P(X > 12)$ **c** $P(X > 6)$ **d** $P(X < 5)$ **e** $P(4 < X < 9)$

*Solution*

**a** $$P(X < 10) = P\left(Z < \frac{10-8}{\sqrt{4}}\right) = P(Z < 1) = \Phi(1) = 0.8413$$

**b** $$P(X > 12) = P\left(Z > \frac{12-8}{\sqrt{4}}\right) = P(Z > 2) = 1 - \Phi(2) = 1 - 0.9772 = 0.0228$$

**c** $$P(X > 6) = P\left(Z > \frac{6-8}{\sqrt{4}}\right) = P(Z > -1) = \Phi(1) = 0.8413$$

**d** $$P(X < 5) = P\left(Z < \frac{5-8}{\sqrt{4}}\right) = P(Z < -1.5) = 1 - \Phi(1.5) = 1 - 0.9332 = 0.0668$$

**e** $$P(4 < X < 9) = P\left(\frac{4-8}{\sqrt{4}} < Z < \frac{9-8}{\sqrt{4}}\right) = P(-2 < Z < 0.5)$$
$$= \Phi(0.5) - \Phi(-2) = 0.6915 - (1 - 0.9772) = 0.6687$$

*Example 3* Find $k$ correct to two decimal places such that **a** $\Phi(k) = 0.9265$
**b** $\Phi(k) = 0.3000$

*Solution*

**a** In Table 3, the tabulated value 0.9265 is located in row 1.4 column .05. We write $k = \Phi^{-1}(0.9265) = 1.45$.

**b** $k = \Phi^{-1}(0.3000) = -\Phi^{-1}(1 - 0.3000) = -\Phi(0.7000)$
0.7000 lies between the tabulated values 0.6985 and 0.7019 in Table 3.
Of these, 0.7000 is closer to 0.6985, so $k = -\Phi^{-1}(0.7000) = -0.52$.

*Example 4* Suppose that $X \sim N(10, 5)$. Find $k$ such that
$P(\mu - k\sigma < X < \mu + k\sigma) = 0.8000$.

*Solution*

$$P(X \le \mu + k\sigma) = P(X < \mu) + P(\mu \le X \le \mu + k\sigma) = 0.5000 + \tfrac{1}{2} \times 0.8000 = 0.9000$$

But

$$P(X \le \mu + k\sigma) = P\left(\frac{X-\mu}{\sigma} \le k\right) = \Phi(k)$$

So

$$\Phi(k) = 0.9000 \Leftrightarrow k = \Phi^{-1}(0.9000) = 1.28$$

A useful characterisation of the Normal distribution is that, whenever $X \sim N(\mu, \sigma^2)$,

$$P(\mu - \sigma < X < \mu + \sigma) = 0.6826$$
$$P(\mu - 2\sigma < X < \mu + 2\sigma) = 0.9544$$
$$P(\mu - 3\sigma < X < \mu + 3\sigma) = 0.9974$$

## EXERCISE 2.7

**1** $Z \sim N(0, 1)$. Use Table 3 (page 139) to find:

**a** $\Phi(0)$ **b** $\Phi(1.00)$ **c** $\Phi(2.00)$
**d** $\Phi(1.38)$ **e** $\Phi(-1.38)$ **f** $\Phi(-2.12)$

**2** $Z \sim N(0, 1)$. Use Table 3 to find:

**a** $P(Z = 1.25)$ **b** $P(Z \geq 0)$ **c** $P(1.00 < Z < 2.15)$
**d** $P(Z > 1.00)$ **e** $P(Z > 2.15)$ **f** $P(-2.15 < Z < -1.00)$
**g** $P(Z > -1.00)$ **h** $P(Z > -2.15)$ **i** $P(-1.00 < Z < 1.00)$
**j** $P(-2.00 < Z < 2.00)$ **k** $P(-3.00 < Z < 3.00)$ **l** $P(-1.96 < Z < 1.96)$

**3** If $X \sim N(20, 25)$, standardise $X$ and use Table 3 to find:

**a** $P(X \leq 25)$ **b** $P(X \leq 15)$ **c** $P(X \geq 25)$
**d** $P(X \geq 15)$ **e** $P(15 \leq X \leq 25)$ **f** $P(10 \leq X \leq 30)$
**g** $P(17 < X < 23)$ **h** $P(16 < X < 28)$ **i** $P(13.5 < X < 23.5)$

**4** $Z \sim N(0, 1)$. Use Table 3 to find:

**a** $\Phi^{-1}(0.9850)$ **b** $\Phi^{-1}(0.6217)$ **c** $\Phi^{-1}(0.7500)$
**d** $\Phi^{-1}(0.025)$ **e** $\Phi^{-1}(0.2000)$ **f** $\Phi^{-1}(0.4298)$

**5** **a** $X \sim N(74, 4)$. Determine the $x$-value with area **(i)** 0.05 to its right **(ii)** 0.10 to its left.
**b** $Y \sim N(335, 100)$. Find the $y$-value with area **(i)** 0.60 to its left **(ii)** 0.60 to its right.
**c** $W \sim N(40.9, 7.1^2)$. Find the three $w$-values that divide the area under the curve into four 0.25 areas. What are these three values called?

**6** A local bottling plant fills bottles of cola which contain 500 ml. The filling machine is set to deliver 503 ml to each bottle. The actual amount put in a bottle is a continuous random variable $X$ where $X \sim N(503, 2.3^2)$. Calculate the probability that a randomly chosen bottle of cola will contain less than the advertised 500 ml.

**7** A machine fills bags with flour whose weight in grams is $W \sim N(\mu, 100)$. The value of $\mu$ can be set by the machine operator. To what value of $\mu$ must the operator set the machine to ensure that 95% of the bags contain at least 1.5 kg?

**8** A farm shop sells free-range hens' eggs. The weights of these eggs are Normally distributed with mean 60 g and standard deviation 10 g. The farm shop classifies eggs as small, medium and large. Any eggs less than 50 g are classified as small. Above what weight should eggs be classed as large if the farm shop wants there to be on average equal numbers of medium and large eggs?

**9** The marks scored by a large number of candidates in an examination are distributed N(55, 225).

**a** What value of $Z \sim N(0, 1)$ would correspond to a score of 76?
It is decided to adjust each candidate's score so that the marks are distributed N(50, 100).
**b** Convert the value you obtained in part **a** to a score on this new scale.
**c** Determine a formula for converting scores on the old scale to scores on the new scale.
**d** Hence, or otherwise, calculate the new score for a candidate whose original mark was **(i)** 55 **(ii)** 34 **(iii)** 82

**10** The speeds of vehicles on a certain stretch of road are thought to be Normally distributed. Based on a large number of observations it is discovered that 80% of vehicles have speeds less than 60 mph and 10% have speeds less than 40 mph. Determine the mean speed $\mu$ and standard deviation $\sigma$.

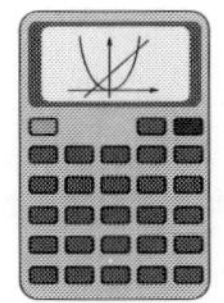

Some calculators will give you the Normal distribution. Look for menu items such as

| | |
|---|---|
| **normalpdf(** | to evaluate the Normal density function |
| **normalcdf(** | to calculate probabilities |
| **invNorm(** | to calculate the value of the random variable for a given probability |

Find out what your calculator can do.

## *Combining Normal random variables*

The laws of expectation and variance have already been demonstrated for discrete random variables. The same rules apply to continuous random variables:

$$E(aX + b) = aE(X) + b$$

$$V(aX + b) = a^2V(X)$$

$$E(X \pm Y) = E(X) \pm E(Y)$$

$$V(X \pm Y) = V(X) + V(Y), \quad \text{provided } X \text{ and } Y \text{ are independent}$$

Furthermore, it can be shown that if $X \sim N(\mu_1, \sigma_1^2)$ and $Y \sim N(\mu_2, \sigma_2^2)$ then

$$X + Y \sim N(\mu_1 + \mu_2, \sigma_1^2 + \sigma_2^2) \quad \text{and} \quad X - Y \sim N(\mu_1 - \mu_2, \sigma_1^2 + \sigma_2^2)$$

provided $X$ and $Y$ are independent.

*Example* $X \sim N(15, 9)$ and $Y \sim N(10, 4)$ are independent random variables.
Calculate: **a** $P(X + Y < 33)$ **b** $P(X > Y)$

*Solution*

**a** $X + Y \sim N(25, 13)$

$$P(X + Y < 33) = P\left(Z < \frac{33 - 25}{\sqrt{13}}\right) = \Phi(2.22) = 0.9868$$

**b** $X - Y \sim N(5, 13)$

$$P(X > Y) = P(X - Y > 0) = P\left(Z > \frac{0 - 5}{\sqrt{13}}\right) = P(Z > -1.39) = \Phi(1.39) = 0.9177$$

## EXERCISE 2.8B

**1** $X \sim N(20, 9)$. Calculate $P(W < 35)$ where $W = 2X - 10$.

**2** $X$ and $Y$ are independent random variables where $X \sim N(10, 9)$ and $Y \sim N(8, 4)$. Calculate:

**a** $P(X + Y < 20)$ **b** $P(X - Y > 0)$

**c** $P(16 < X + Y < 24)$ **d** $P(-1 < X - Y < 3)$

**3** $X$ and $Y$ are independent random variables where $X \sim N(5, 0.25)$ and $Y \sim N(4, 0.64)$. Calculate $P(2X - 3Y > 0)$.

**4** Use the laws of expectation and variance to show that if $X \sim N(\mu, \sigma^2)$ then

**a** $E\left(\frac{X - \mu}{\sigma}\right) = 0$ **b** $V\left(\frac{X - \mu}{\sigma}\right) = 1$

**5** The maximum load a certain motorbike can carry is 140 kg. The weights of men and women (in kilograms) are assumed to be distributed N(65, 100) and N(55, 64).

**a** If a randomly chosen man and woman ride the motorbike, what is the probability that it would be overloaded? What assumption(s) have you made?

**b** The manufacturers of the motorbike want to be 99% certain that it is not overloaded when the motorbike is ridden by two men chosen at random. What should the motorbike's maximum load be (to the nearest kg)?

**6** It can be assumed that the birth weights (in kilograms) of male and female babies are distributed N(3.6, 0.16) and N(3.3, 0.09) respectively.

**a** Find the probability that, of a male baby and a female baby selected at random, the male is heavier.

**b** One day, in a large hospital, eight females and ten males are born. What is the distribution of their total birth weight?

**7** During safety trials of a new drug, 36 rats are to be injected with the drug at a dose level of 3 units per gram of body weight. The body weight of these rats is distributed N(30, 16).

**a** State the distribution of the number of units of drug required for a randomly selected rat.

**b** What is the distribution of the number of units of drug required for the whole experiment?

**c** If the experimenter has 3000 units of the drug available, what is the probability that there is not enough to complete the experiment?

**8** A manufacturing process involves fitting two components A and B inside a slot C as shown.

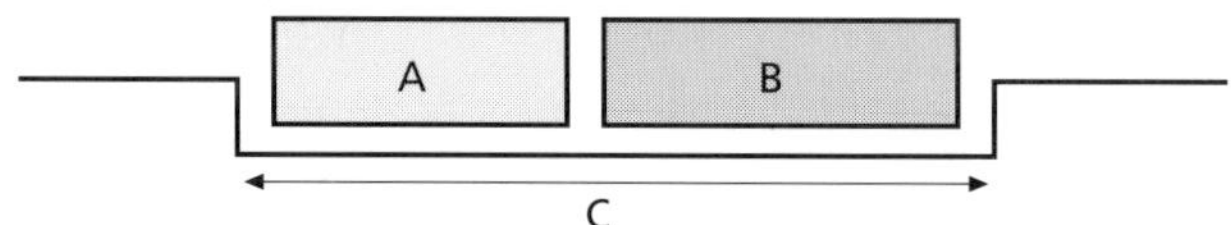


The lengths (in mm) of the components A and B are independently distributed N(10, 1) and N(12, 1) respectively, and the lengths of slot C are distributed N(24, 2). What is the probability that randomly selected components A and B will not fit into one of these slots?

## Approximating the Binomial with the Normal

To simplify the calculations, we can sometimes use the Normal distribution in situations where the Binomial is the appropriate model but $n$ is so large that the arithmetic becomes cumbersome.

> If $X \sim \text{Bin}(n, p)$, $n$ large and $p$ close to 0.5, then $X \sim \text{N}(np, npq)$ approximately.

*Rule of thumb*

Use the Normal approximation provided both $np$ and $nq$ are greater than 5. Remember that the Poisson gives a good approximation to the Binomial when $n$ is large and $p$ is small. If neither approximation can be used, then Binomial probabilities must be calculated from the usual formula.

For example, suppose $X \sim \text{Bin}(10, 0.5)$. Notice that, in this case, like the Normal, the distribution has a symmetrical shape.

The probability distribution Bin(10, 0.5)

From tables, $\text{P}(X = 6) = 0.2051$.

$$\text{E}(X) = np = 10 \times 0.5 = 5 \text{ and}$$

$$\text{V}(X) = npq = 10 \times 0.5 \times 0.5 = 2.5$$

so the Normal approximation is $Y \sim \text{N}(5, 2.5)$.
Now, with a continuous distribution, $\text{P}(Y = y) = 0$ so $\text{P}(Y = 6) = 0$.

However, if we think of the probability spread over the interval from 5.5 to 6.5 we have

$$\text{P}(5.5 < Y < 6.5) = \text{P}\left(\frac{5.5. - 5}{\sqrt{2.5}} < Z < \frac{6.5 - 5}{\sqrt{2.5}}\right)$$

$$= \text{P}(0.32 < Z < 0.95)$$

$$= \Phi(0.95) - \Phi(0.32) = 0.8289 - 0.6255 = 0.2034$$

which is quite close to 0.2051, the answer from Binomial tables.

As $n$ increases the approximation improves. Even when $p$ is not close to 0.5, if $n$ is large enough the approximation is satisfactory provided $np$ and $nq$ are both greater than 5. For example, the following diagrams show that, as $n$ increases, the shape of the Binomial distribution $\text{Bin}(n, 0.1)$ becomes more symmetrical and so more suitable for approximation by the Normal.

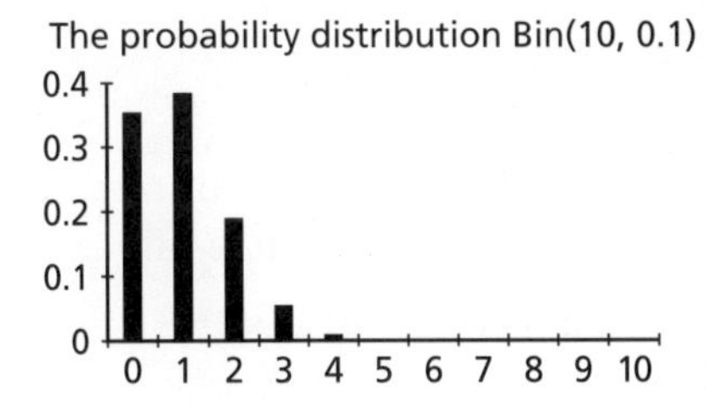

The probability distribution Bin(10, 0.1)

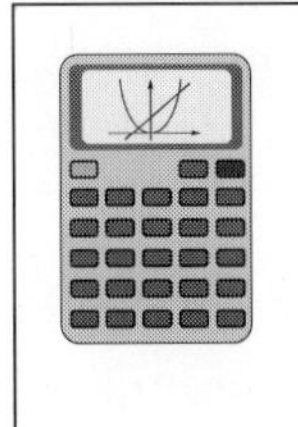

Some calculators will draw bar diagrams or histograms.
Find out what your calculator can do.

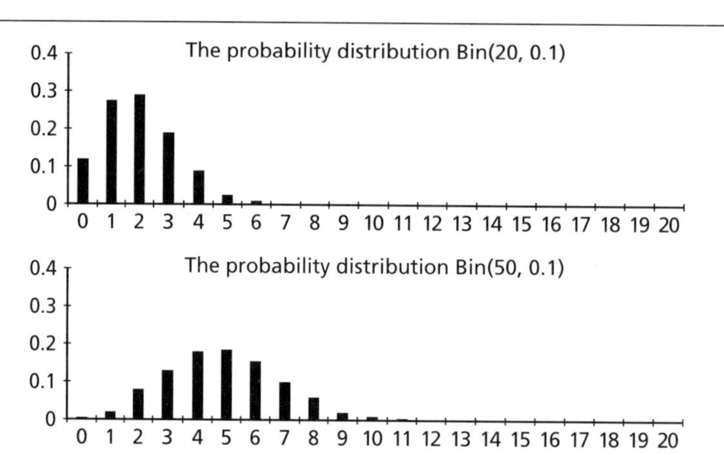


Remember that, because $X \sim \text{Bin}(n, p)$ is discrete while the Normal is a model for continuous random variables, we can not evaluate $P(X = x)$ directly when using the approximation $Y \sim N(np, npq)$. Instead we have to think of the probability as being spread over the interval $x - 0.5 < Y < x + 0.5$.

This is what is known as a **continuity correction**:
if $X \sim \text{Bin}(n, p)$ and $Y \sim N(np, npq)$

$P(X < x)$ becomes $P(Y < x - 0.5)$
$P(X \le x)$ becomes $P(Y < x + 0.5)$
$P(X \ge x)$ becomes $P(Y > x - 0.5)$
$P(X > x)$ becomes $P(Y > x + 0.5)$

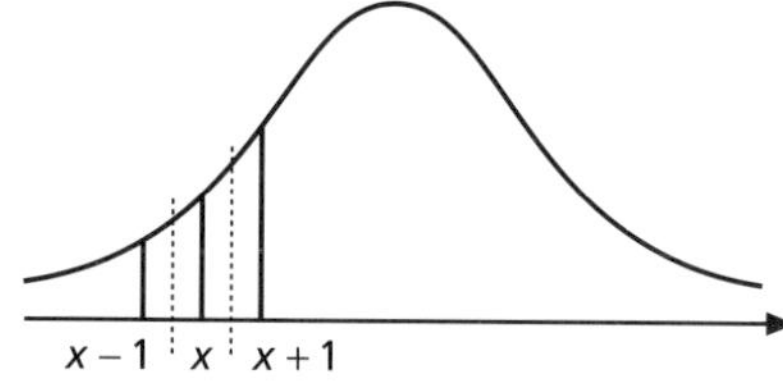


The boundary between consecutive values of the discrete random variable is taken at the half-way mark.

*Example* $X \sim \text{Bin}(16, 0.4)$. Calculate:
**a** $P(X \le 6)$ **b** $P(X > 8)$ using the Normal approximation and compare your answers with the values in Table 1 (page 135).

*Solution*
$np = 16 \times 0.4 = 6.4$ and $nq = 16 \times 0.6 = 9.6$
Both $np$ and $nq$ are greater than 5 so use the Normal approximation.

$$E(X) = np = 6.4 \quad \text{and} \quad V(X) = npq = 16 \times 0.4 \times 0.6 = 3.84, \quad \text{so } Y \sim N(6.4, 3.84)$$

**a** $$P(X \le 6) = P(Y < 6.5) = P\left(Z < \frac{6.5 - 6.4}{\sqrt{3.84}}\right) = \Phi(0.05) = 0.5199$$

From Table 1, $P(X \le 6) = 0.5272$.

**b** $$P(X > 8) = P(Y > 8.5) = P\left(Z > \frac{8.5 - 6.4}{\sqrt{3.84}}\right) = P(Z > 1.07)$$
$$= 1 - \Phi(1.07) = 1 - 0.8577 = 0.1423$$

From Table 1, $P(X > 8) = 1 - P(x \le 8) = 1 - 0.8577 = 0.1423$.

The Normal distribution gives a reasonable approximation to the Binomial in this case. The Normal gives a very good approximation to the Binomial when $n$ is larger.

The Normal approximation to the Binomial is a special case of the Central Limit Theorem which we shall meet in the next chapter.

## EXERCISE 2.9B

1 $X \sim \text{Bin}(20, 0.5)$ Calculate $P(X \leq 10)$ using
  a Binomial tables,
  b the Normal approximation (i) with (ii) without a continuity correction.

2 a $X \sim \text{Bin}(100, 0.5)$ Find $P(X \leq 50)$ (i) with (ii) without a continuity correction.
  b $X \sim \text{Bin}(1000, 0.5)$ Find $P(X \leq 500)$ (i) with (ii) without a continuity correction.
  c How does the effect of the continuity correction change as $n$ increases?

3 Thirty fair coins are tossed at the same time and $X$ is the number of Heads which appear.
  a State the distribution of $X$ and calculate the mean and standard deviation of $X$.
  b Use the Normal approximation to calculate:
    (i) $P(X < 18)$ (ii) $P(X \leq 18)$ (iii) $P(X > 14)$ (iv) $P(X \geq 14)$

4 Assuming that 20% of people are left-handed, use a suitable approximation to calculate the probability that there will be more than 10 left-handed pupils among 80 sixth-year pupils.

5 A manufacturer sells a particular type of industrial fastener in boxes of 100. The proportion of defective fasteners produced by the factory is 0.8%. Using a suitable approximation, calculate the probability that a box contains at most one defective fastener.

6 Airline overbooking is common practice. Past records show that 20% of people making a reservation fail to show up. An airline operates a shuttle service between two cities using aircraft that have 200 seats. They regularly accept 230 reservations for each flight. Let $X$ be the random variable representing the number of passengers who show up.
  a Assuming passengers act independently, state the distribution of $X$.
  b Calculate approximately the probability that, on any given flight, at least one passenger holding a reservation will not have a seat.
  c Why might the assumption of independence be unrealistic?

7 Spillages of diesel fuel on the road are environmentally harmful and can cause vehicles to skid and lose control. It is suspected that 10% of commercial vehicles with diesel engines have fuel tanks which leak. As part of a coordinated nationwide campaign, police forces set up road blocks at various points across the country one day and 100 commercial vehicles were stopped and inspected at each site. Calculate the approximate probability that a randomly selected road block found fifteen or more vehicles with defective fuel tanks.

# STATISTICS IN ACTION – MODELLING DATA

## *1. True / false quizzes*

Have you ever sat a multiple-choice exam and wondered how many marks are awarded to candidates who simply guess the correct answer? Psychologists and examiners have developed theories of mental testing which attempt to address this and other issues. To gain some insight into how people behave on tests you might like to try the following experiment, which uses a very simple type of mental test, namely a true/false quiz.

Make up a short list of statements that might be true or false. The idea is to think of statements which seem to be straightforward general knowledge but which are sufficiently obscure that it is most likely that those taking the quiz will *guess* true or false. A light-hearted and entertaining tone to the list is recommended and six to eight statements each taken from a different area of knowledge should be sufficient. Choose statements suitable for the people you are going to ask to take part in the experiment.

One such quiz consisted of the following seven statements.

1. When exposed to a wind of 30 miles per hour in a temperature of –30 °F, human flesh freezes solid in 30 seconds.
2. A coho is a fish.
3. Omdurman is a town in Morocco.
4. The probability of getting a royal flush in poker is about 1 in 650 000 hands.
5. A brool is a deep murmur.
6. Before anaesthetics were invented, the shortest time recorded for a leg amputation was 20 seconds by Napoleon's chief surgeon, Dominic Larrey.
7. The fonticulus is a little dip just on the top of the breast bone.

The answers are:
1. True   2. True   3. False (it is in Sudan)   4. True   5. True
6. False (he took 13–15 seconds)   7. True

The discrete random variable $X$ is the number of correct responses achieved by a contestant and the range space for $X$ is $\{0, 1, 2, 3, 4, 5, 6, 7\}$. If we assume that each contestant is guessing, the probability of a correct response to each of the questions is 0.5. This assumption will be violated if the contestant actually knows the correct answer. We will assume that each contestant is acting independently, i.e. is not discussing possible answers with others or copying others' responses and that their response to each question is not influenced by their responses to any of the other questions. A possible probability model is

$$X \sim \text{Bin}(7, 0.5)$$

| Score $x$ | 0 | 1 | 2 | 3 | 4 | 5 | 6 | 7 |
|---|---|---|---|---|---|---|---|---|
| $P(X = x)$ | 0.0078 | 0.0547 | 0.1641 | 0.2734 | 0.2734 | 0.1641 | 0.0547 | 0.0078 |

If a total of 107 pupils took the above quiz, we would expect to see about these same proportions in the sample, which means we would expect about the following number of pupils in each category:

| Score x | 0 | 1 | 2 | 3 | 4 | 5 | 6 | 7 | Total |
|---|---|---|---|---|---|---|---|---|---|
| Expected frequency $E$ | 0.8 | 5.9 | 17.6 | 29.3 | 29.3 | 17.6 | 5.9 | 0.8 | 107.2 |

where $E = P(X = x) \times 107$

The following table summarises the results of the quiz for 107 second-year pupils. As well as the observed frequency for each value of $X$, the expected frequency is shown for comparison.

| Score x | 0 | 1 | 2 | 3 | 4 | 5 | 6 | 7 | Total |
|---|---|---|---|---|---|---|---|---|---|
| Observed frequency $O$ | 0 | 8 | 25 | 31 | 20 | 18 | 5 | 0 | 107 |
| Expected frequency $E$ | 0.8 | 5.9 | 17.6 | 29.3 | 29.3 | 17.6 | 5.9 | 0.8 | 107.2 |

The slight difference between the two totals is because of rounding errors.

A comparison of the experimental frequencies observed with the theoretical frequencies predicted by the Binomial model is illustrated in the graph.

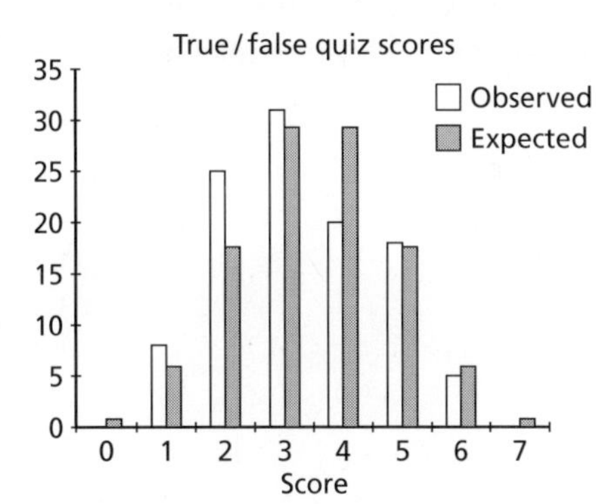


The model seems to fit the data reasonably well, although we note the discrepancies when $X = 2$ and $X = 4$.

In Statistics 2 (AH), a method of quantifying and assessing the differences between observed and expected frequencies is studied.

Collect your own data and analyse it in a similar way.

## *2. Detecting background radiation*

Using a Geiger counter borrowed from the school's Physics Department the number of particles detected as background radiation per 10 second interval was recorded for a total of 30 minutes. The data are shown here and may be read across rows as a time series.

| | | | | | | | | | | | | | | | | | |
|---|---|---|---|---|---|---|---|---|---|---|---|---|---|---|---|---|---|
| 6 | 6 | 3 | 5 | 10 | 5 | 3 | 4 | 7 | 4 | 5 | 6 | 3 | 7 | 6 | 6 | 2 | 5 |
| 11 | 3 | 6 | 5 | 3 | 5 | 4 | 8 | 6 | 1 | 10 | 6 | 3 | 4 | 2 | 5 | 9 | 2 |
| 3 | 3 | 5 | 4 | 5 | 9 | 5 | 4 | 8 | 5 | 5 | 5 | 2 | 7 | 3 | 6 | 1 | 6 |
| 9 | 4 | 4 | 8 | 5 | 5 | 1 | 6 | 7 | 2 | 6 | 4 | 7 | 6 | 4 | 2 | 3 | 8 |
| 4 | 6 | 3 | 0 | 3 | 5 | 7 | 6 | 7 | 5 | 8 | 5 | 5 | 4 | 8 | 7 | 3 | 5 |
| 6 | 6 | 8 | 11 | 5 | 5 | 3 | 6 | 6 | 8 | 4 | 6 | 3 | 3 | 6 | 6 | 7 | 5 |
| 3 | 11 | 8 | 5 | 3 | 7 | 6 | 4 | 4 | 6 | 8 | 5 | 6 | 3 | 3 | 9 | 4 | 6 |
| 2 | 6 | 5 | 8 | 4 | 0 | 6 | 7 | 7 | 5 | 4 | 9 | 7 | 5 | 4 | 7 | 10 | 2 |
| 5 | 4 | 4 | 7 | 2 | 4 | 8 | 6 | 3 | 2 | 4 | 10 | 6 | 5 | 5 | 8 | 2 | 5 |
| 3 | 5 | 5 | 6 | 2 | 3 | 11 | 3 | 5 | 5 | 5 | 5 | 4 | 6 | 3 | 5 | 12 | 7 |

Construct an ungrouped frequency table for these data.

Let the discrete random variable $X$ be the number of particles detected per 10 second interval. Suppose that these data come from a Poi($\mu$) distribution, where $\mu$ is the theoretical mean count in a 10 second interval. Calculate the sample mean $\bar{x}$.

Using $\bar{x}$ as though it were the true value of $\mu$, calculate $P(X = x)$ for the following values of $X$:

$$\{0, 1, 2, 3, 4, 5, 6, 7, 8, 9, 10, 11, 12 \text{ or more}\}$$

Calculate expected frequencies $E$ for these values of $X$ where $E = P(X = x) \times 180$.

Draw a bar diagram similar to the one on page 67 and compare the observed and expected frequencies. Comment on the adequacy of the Poisson model you used for these data.

Use a Geiger counter to collect some data on the background radiation in your classroom. Analyse your results in a similar way and compare with the above data.

### *3. Are heights Normally distributed?*

The probability density function for the Normal distribution has a characteristic 'bell' shape. When we illustrate a sample of heights using a histogram or stem and leaf diagram, it is sometimes difficult to recognise this shape.

Heights of S6 male students

| | Stem | Leaf |
|---|---|---|
| | 16 | 6 7 |
| | 17 | 0 2 2 3 3 3 4 |
| | 17 | 5 5 6 6 7 8 9 9 |
| | 18 | 0 0 1 2 2 3 4 |
| | 18 | 6 6 7 8 9 9 |
| | 19 | 2 |
| $n = 31$ | 16 | 6 represents 166 cm |

For example, this stem and leaf diagram shows the heights of 31 male students in S6.

With small samples there is insufficient data to get a clear picture of the structure of the variation observed. All we can say is that these data could have come from a Normally distributed population.

An alternative method for examining sample data is to construct a **probability plot**. First, sort the data into ascending order $y_{(1)} \leq y_{(2)} \leq \ldots \leq y_{(i)} \leq \ldots \leq y_{(n)}$. In our example, this is achieved by the stem and leaf diagram.

Then calculate

$$z_i = \Phi^{-1}\left(\frac{i}{n+1}\right) \text{ for } i = 1, 2, \ldots, n \quad \text{where } Z \sim N(0, 1)$$

The $z_i$ are called **Normal scores**. In our example

$$z_1 = \Phi^{-1}\left(\frac{1}{31+1}\right) = \Phi^{-1}(0.03125) = -\Phi^{-1}(1 - 0.03125) = -\Phi(0.96875) = -1.86$$

$$z_2 = \Phi^{-1}\left(\frac{2}{31+1}\right) = \Phi^{-1}(0.0625) = -\Phi^{-1}(1 - 0.0625) = -\Phi^{-1}(0.9375) = -1.53$$

$$z_3 = \Phi^{-1}\left(\frac{3}{31+1}\right) = \Phi^{-1}(0.09375) = -\Phi^{-1}(1 - 0.09375) = -\Phi^{-1}(0.90625) = -1.32$$

$$\vdots$$

$$z_{15} = \Phi^{-1}\left(\frac{15}{31+1}\right) = \Phi^{-1}(0.46875) = -\Phi(1 - 0.46875) = -\Phi^{-1}(0.53125) = -0.08$$

$$z_{16} = \Phi^{-1}\left(\frac{16}{31+1}\right) = \Phi^{-1}(0.5) = 0.00$$

$$z_{17} = \Phi^{-1}\left(\frac{17}{31+1}\right) = \Phi^{-1}(0.53125) = 0.08$$

$$\vdots$$

$$z_{31} = \Phi^{-1}\left(\frac{31}{31+1}\right) = \Phi^{-1}(0.96875) = 1.86$$

Finally, plot the points $(z_i, y_{(i)})$.

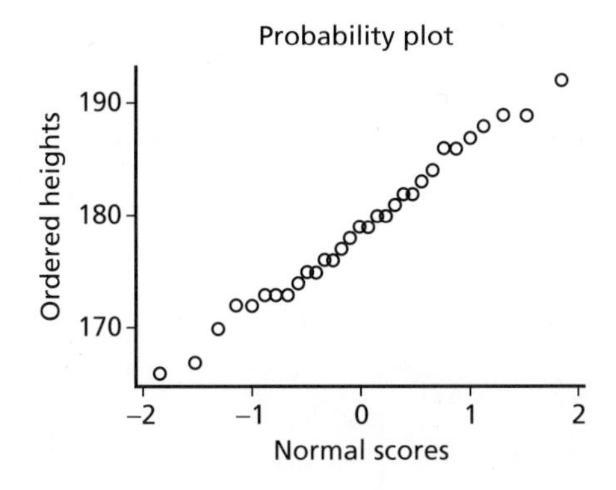


In our example, the resulting plot shows the points lying roughly in a straight line so we conclude that the sample data are from a Normal population.

Measure the heights of a random sample of nine female S6 students and construct a probability plot. Is it plausible that the heights of female students in S6 are Normally distributed?

Some calculators will draw Normal probability plots.
Look for menu items such as **NormProbPlot**.
Find out what your calculator can do.

# CHAPTER 2 SUMMARY

**1** **(i)** A **random variable** is a function which associates a unique real value with each outcome in the sample space of a random experiment.
**(ii)** If the **range space** of the random variable $X$ can be written in the form $\{x_1, x_2, x_3, \ldots\}$ then we say that $X$ is a **discrete random variable**.
**(iii)** With each value $x$ of the discrete random variable $X$ we may associate the probability of its occurrence $P(X = x) = p(x)$.

**2** The set of all pairs $(x, p(x))$ is called the **probability distribution** of $X$. Note that

$$0 \le p(x) \le 1$$

and $$\sum_{\text{all } x} p(x) = 1$$

**3** The **mean** or **expected value** of $X$ is

$$\mu = E(X) = \sum_{\text{all } x} x\,p(x)$$

**4** The **variance** of $X$ is

$$\sigma^2 = V(X) = E(X^2) - (E(X))^2$$

where $$E(X^2) = \sum_{\text{all } x} x^2 p(x)$$

**5** **Laws of expectation and variance**

$$E(aX + b) = aE(X) + b$$
$$V(aX + b) = a^2 V(X)$$
$$E(X \pm Y) = E(X) \pm E(Y)$$
$V(X \pm Y) = V(X) + V(Y)$, provided $X$ and $Y$ are independent

**6** **The Binomial distribution**
The random variable $X$ is said to follow a binomial distribution i.e. $X \sim \text{Bin}(n, p)$ if

$$P(X = x) = \begin{cases} \binom{n}{x} p^x q^{n-x} & x = 0, 1, 2, \ldots, n \\ 0 & \text{otherwise} \end{cases}$$

where $0 < p < 1$ and $q = 1 - p$.

$$E(X) = np \quad \text{and} \quad V(X) = npq$$

The required conditions are:
**(i)** there is a fixed number of $n$ trials;
**(ii)** only two outcomes, ‘success’ or ‘failure’, are possible at each trial;
**(iii)** the trials are independent;
**(iv)** there is a constant probability $p$ of success;
**(v)** the random variable is the total number of successes in $n$ trials.

**7 The Poisson distribution**

The random variable $X$ follows a Poisson distribution, i.e. $X \sim \text{Poi}(\mu)$ if

$$P(X = x) = \frac{e^{-\mu}\mu^x}{x!} \quad \text{for } x = 0, 1, 2, 3, \ldots$$

$$E(X) = V(X) = \mu$$

The required conditions are:

**(i)** isolated events are occurring in continuous time or space;

**(ii)** the numbers of events that occur in non-overlapping segments of time or space are independent;

**(iii)** events occur singly rather than in groups;

**(iv)** events are occurring at a constant average rate per unit time or space throughout the whole period or region of interest;

**(v)** the random variable is the numbers of events occurring per segment.

**8 Poisson approximation to the Binomial**

If $n$ is large (say $n \geq 20$) and $p$ is small (say $p \leq 0.05$) then $\text{Poi}(np)$ provides a good approximation to $\text{Bin}(n, p)$. The approximation is very good when $n \geq 100$ and $np < 10$.

**9 Continuous random variables**

**(i)** Random experiments involving measurement result in real number values being recorded and the random variable can take any value in a range of values. It is not possible to assign probabilities to individual values of a **continuous random variable** and $P(X = x) = 0$. Instead, probabilities are determined for an interval of values of $X$ by calculating areas under the **probability density function** of $X$. The shape of the pdf shows how probability is spread across the range of values of $X$.

**(ii)** If the continuous random variable $X$ has the pdf $f(x)$ defined on the interval $a < x < b$, then the **cumulative distribution function** of $X$ is

$$F(x) = P(X \leq x) = \int_a^x f(u)\,du$$

and

$$P(x_1 < X < x_2) = F(x_2) - F(x_1)$$

Alternatively

$$P(x_1 < X < x_2) = \int_{x_1}^{x_2} f(x)\,dx$$

where $a < x_1 < x_2 < b$.

**(iii)** The pdf $f(x) = \frac{d}{dx}F(x)$

$$\mu = E(X) = \int_a^b x f(x)\,dx$$

$$\sigma^2 = V(X) = E(X^2) - \mu^2$$

where

$$E(X^2) = \int_a^b x^2 f(x)\,dx$$

**10 The Normal distribution**

**(i)** A continuous random variable $X$ is said to have a Normal distribution if it has the probability density function

$$f(x) = \frac{1}{\sqrt{2\pi\sigma^2}} \exp\left(-\frac{(x-\mu)^2}{2\sigma^2}\right), \quad -\infty < x < \infty$$

We write $X \sim \mathrm{N}(\mu, \sigma^2)$.

**(ii)** The Normal distribution has two parameters

$$\mu = \mathrm{E}(X) \quad \text{and} \quad \sigma^2 = \mathrm{V}(X)$$

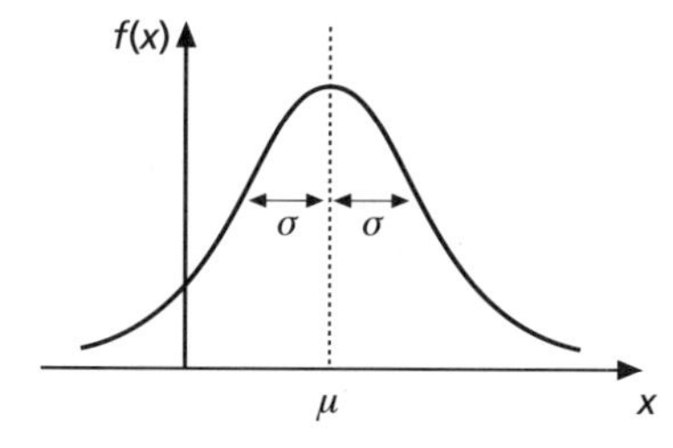


**(iii)** The random variable $X \sim \mathrm{N}(\mu, \sigma^2)$ can be transformed to the **Standard Normal** random variable $Z \sim \mathrm{N}(0, 1)$ where $Z = \frac{X-\mu}{\sigma}$ so that

$$\mathrm{P}(X < x) = \mathrm{P}\left(Z < \frac{x-\mu}{\sigma}\right) = \Phi\left(\frac{x-\mu}{\sigma}\right)$$

Tables of the cumulative Standard Normal distribution $\Phi(z)$ can then be used to evaluate any Normal probability.

**11 Combining Normal random variables**

If $X \sim \mathrm{N}(\mu_1, \sigma_1^2)$ and $Y \sim \mathrm{N}(\mu_2, \sigma_2^2)$, then

$$X + Y \sim \mathrm{N}(\mu_1 + \mu_2, \sigma_1^2 + \sigma_2^2)$$

and

$$X - Y \sim \mathrm{N}(\mu_1 - \mu_2, \sigma_1^2 + \sigma_2^2)$$

provided $X$ and $Y$ are independent.

**12 Normal approximation to the Binomial**

**(i)** If $X \sim \mathrm{Bin}(n, p)$, for $n$ large and $p$ close to 0.5, then $X \sim \mathrm{N}(np, npq)$ approximately.

**(ii)** *Rule of thumb*: use the Normal approximation provided both $np$ and $nq$ are greater than 5.

**(iii)** Because $X \sim \mathrm{Bin}(n, p)$ is discrete while the Normal is a model for continuous random variables, a **continuity correction** should be applied.
If $X \sim \mathrm{Bin}(n, p)$ is approximated by $Y \sim \mathrm{N}(np, npq)$ then

$\mathrm{P}(X = x)$ becomes $\mathrm{P}(x - 0.5 < Y < x + 0.5)$
$\mathrm{P}(X < x)$ becomes $\mathrm{P}(Y < x - 0.5)$
$\mathrm{P}(X \leq x)$ becomes $\mathrm{P}(Y < x + 0.5)$
$\mathrm{P}(X \geq x)$ becomes $\mathrm{P}(Y > x - 0.5)$
$\mathrm{P}(X > x)$ becomes $\mathrm{P}(Y > x + 0.5)$

# CHAPTER 2 REVIEW EXERCISE



**1** The number of spots on opposite faces of a fair die add up to 7. A dishonest gambler changed a die so that the face opposite the 6 also has six spots (instead of one spot). The discrete random variable $X$ represents the score produced by such a die. Determine the probability distribution of $X$. Calculate the mean and variance of $X$.

**2** The random variable $X$ has a mean of 10 and a variance of 4. The random variable $Y$ has a mean 20 and standard deviation 3. The random variables $X$ and $Y$ are independent. Calculate the mean and variance of:

**a** $X + Y$ **b** $X - Y$ **c** $5X - 2Y$ **d** $2X + Y - 15$

**3** $X \sim \text{Bin}(7, 0.3)$. Calculate:

**a** $P(X = 3)$ **b** $P(X = 5)$ **c** $P(X \leq 2)$ **d** $P(X > 2)$

**4** A multiple-choice test consists of twelve questions and there are five responses to each question (of which only one is correct). A student guesses the answer to each question.

**a** What is the probability he gets exactly six correct answers?

**b** Calculate the mean and standard deviation of the number of correct answers he achieves.

**5** $X \sim \text{Poi}(0.7)$. Calculate the first four values of the probability distribution of $X$.

**6** On average, cars cross a level-crossing at a rate of 90 per hour. When a train passes, the level-crossing gates are closed for three minutes. What is the probability that there are at most two cars waiting when the gates open after the train has passed?

**7** On average, 2% of the items produced on a production line are defective. The items are packed in boxes of 100. Use the Poisson approximation to the Binomial to calculate the probability that a box contains:

**a** no defective items,

**b** one defective item,

**c** at most two defective items.

**8** The continuous random variable $X$ has probability density function

$$f(x) = \begin{cases} k(x-1)(x-5) & 1 < x < 5 \\ 0 & \text{otherwise} \end{cases}$$

Determine:

**a** the value of $k$,

**b** the cumulative distribution function,

**c** $P(2 < X < 4)$.

**9** $Z \sim N(0, 1)$. Calculate:

**a** $\Phi(1.33)$ **b** $\Phi(-2.18)$ **c** $\Phi^{-1}(0.9484)$ **d** $\Phi^{-1}(0.3300)$

**10** $X \sim \mathrm{N}(9, 4)$. Calculate:

**a** $\mathrm{P}(X < 5)$ **b** $\mathrm{P}(X > 10)$ **c** $\mathrm{P}(6 \leq X \leq 12)$

**11** A company buys steel rods of nominal length 10 m. The actual length has a Normal distribution N(9.8, 0.04). Rods less than 9.9 m result in less profit for the company. In a consignment of 1000 rods, what is the expected number of rods less than 9.9 m?

**12** The quadrilateral ABCD is nominally a rectangle of size 12 cm by 6 cm. Actually, AB and CD are independently distributed N(12, 4), while BC and AD are independently distributed N(6, 1). What is the probability that the perimeter is less than 35 cm?

**13** The heights (in cm) of S1 boys and girls are assumed to be distributed $\mathrm{N}(155, 10^2)$ and $\mathrm{N}(148, 9^2)$ respectively. Calculate the probability that, for a randomly chosen boy and girl, the girl is taller.

**14** The probability of successfully treating a certain disease is 0.6. If 200 patients are treated, what is the approximate probability that more than 70% of them will be cured?

# 3 Sampling



## Populations and samples

What proportion of Scottish voters intend to vote SNP (Scottish National Party) at the next Election for the Scottish Parliament? Most research is carried out to find an answer to a question, like this, which refers to a particular **population**. Often, the question focuses on one **parameter**, or feature of the population, in this case the population proportion of voters who intend to vote SNP.

If we could get an honest response from every Scottish voter, then we would know the answer to our question exactly. We would know the value of the population parameter exactly. A study in which data are obtained from every member of a population is called a **census**. Most research, however, is conducted on a small **sample** rather than the whole population. In this context, the most important reason for sampling would be cost.

- **Cost**. Sampling greatly reduces the cost of collecting data. For example, an opinion poll is usually conducted by surveying the views of between 1000 and 3000 voters, while a General Election records the votes of millions of people.

There are other reasons for sampling, which might be more important in other contexts.

- **Speed**. The results of a sample can be processed much more quickly than those from a census or complete enumeration. For example, the Retail Price Index has to be published every month and this tight time schedule requires the index to be based on a sample of prices in retail outlets.
- **Accuracy**. Surprisingly, the results from a sample can be more accurate than those from a complete enumeration or census. For example, when the accounts of a large business are being audited, a sample of the accounts or invoices will be examined. This means that fewer staff are required and they can be more readily trained and supervised to work to uniform standards. Furthermore, the smaller volume of work means that staff maintain their interest throughout the project and thus make fewer mistakes.
- **Hazard**. Some investigations result in members of the population being endangered or destroyed. In such cases it is not feasible or ethical to test the whole population. For example, a new drug might have unpleasant side effects, so it is tested under carefully controlled conditions on a sample of patients before it is authorised for general use.
- **Accessibility**. Some members of the population may not be observable at all and so we are compelled to work with a sample rather than the whole population. For example, historical records made during a period of interest may be incomplete and so we are forced to limit our examination of them to those which are available.

Although sampling has many advantages, it also has a major disadvantage. Because not all the members of the population are observed directly, the original question can't be answered with complete certainty. For example, we might discover that 420 out of a sample of 1000 Scottish voters intend to vote SNP. Although we might assume that this sample proportion of 42% is a good estimate of the population proportion, we would not know the population parameter exactly.

**Statistical inference** is the study of how to use information from a sample to draw valid conclusions about the population from which the sample was drawn. Study of the sample may produce a **sample statistic** such as the sample mean $\overline{x}$ which is used to **infer** the value of the corresponding **population parameter** namely the population mean $\mu$.

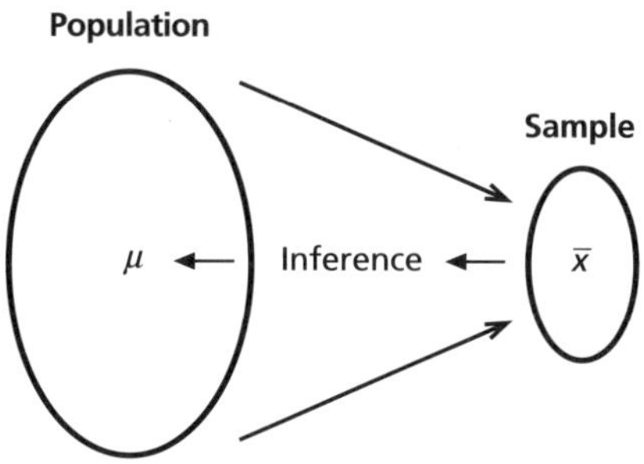

A sample should be **representative** of the population from which it is drawn. We want a sample with the same characteristics as the population. A method of sampling which does not result in an honest representation of the population is said to be **biased**. This can result in wrong conclusions being drawn about the population.

### *Some definitions*

**Population** — the whole collection of individuals (people, animals, plants or things) that we aim to study and about which we intend to draw conclusions.

**Sample** — the subset of the population from which we intend to collect data. We try to ensure that the sample gives an honest representation of the population.

**Sampling units** — non-overlapping collections of individuals from the population. In simple cases, the units are individual members of the population. However, sometimes it is convenient to define larger units such as households, post-code districts, schools, etc.

**Sampling frame** — an exhaustive list of all the units in the population. Care must be taken to avoid incompleteness, duplication and inaccuracy.

**Sampling fraction** — the proportion of the population that is sampled. If a sample of size $n$ is drawn from a population of size $N$ then the sampling fraction is $\frac{n}{N}$.

For example, suppose we wish to estimate the number of children in families that live in a particular housing estate. The sampling unit is a household. (Note that in the context of this investigation our concept of the term 'household' may be challenged.) The population is all the households on the estate. The sampling frame could be the electoral register, a list of postal addresses or a list of addresses for council tax collection.

## EXERCISE 3.1

**1** List five reasons why a small sample may be studied rather than the whole population.

**2** What qualities should a sample have?

**3** Describe a major disadvantage of sampling.

**4** A sample survey is undertaken to estimate the mean height of first-year girls in a large secondary school. Describe:
**a** the sampling units, **b** the population, **c** the sampling frame.

**5** A School Board is discussing the school's uniform policy and wishes to consider the views of parents. To reduce cost, it decides to conduct a sample survey. In this context, describe:
**a** the sampling units, **b** the population, **c** the sampling frame.

**6** **a** How might you conduct a sample survey to estimate the mean number of chairs per classroom in your school?
**b** Describe:
**(i)** the sampling units, **(ii)** the population, **(iii)** the sampling frame.

**7** **a** How might you conduct a sample survey to estimate the average number of times per year that a book is borrowed from the school library?
**b** In this context, describe:
**(i)** the sampling units, **(ii)** the population, **(iii)** the sampling frame.

**8** Describe how you might conduct a sample survey to answer each of the following questions:
**a** How many TVs are there per family amongst pupils at your school?
**b** What proportion of households in a particular town have more than one car?
**c** What is the total amount of seaweed on a particular beach after a storm?
In each case identify
**(i)** the sampling units, **(ii)** the population, **(iii)** the sampling frame.

## *Simple random sampling*

**Probability sampling** is a general name given to a variety of methods where, to avoid selection bias, a random process is used to choose the units to include in the sample. The probability that any unit in the population will be sampled is known beforehand, although this probability need not be the same for all units. All the formal methods of statistical inference assume probability sampling.

**Simple random sampling** is the most basic form of probability sampling. Units are selected for the sample by numbering all the units in the sampling frame consecutively, from 1 to $N$ say, and then choosing $n$ of them randomly. This means that every unit in the population has the same chance of being sampled. The random selection process can be achieved in a number of ways.

- **A lottery**. Place $N$ tickets numbered 1 to $N$ in a container, mix them thoroughly, and pick out $n$ of them without replacement. Choose those units in the sampling frame with the same numbers as on the tickets drawn from the container. A problem with this method is that it is very difficult to ensure that the numbered tickets are thoroughly mixed before selection begins.
- **Using a calculator**. Generate $n$ distinct random numbers in the range 1 to $N$ using a calculator. Choose those units in the sampling frame with the same numbers as produced by the calculator.

  Most calculators have a button or function which produces random numbers between 0 and 1. For example

  $$Rand = 0.329, 0.726, 0.241, \text{ and so on}$$

  These can be converted to whole numbers in the range 1 to $N$ using the formula

  $$Int(Rand \times N) + 1$$

  *Int*() means 'take the integer part of the number between the brackets'. Some calculators have a single command that will produce the required numbers. For example

  $$randInt(1, N)$$

  Find out what your calculator can do.
- **Using a spreadsheet**. Spreadsheets can be used to generate a list of random numbers in the range 1 to $N$ which could be used to select the units to include in the sample. Alternatively, if the sampling frame is listed in a spreadsheet where each row corresponds to a sampling unit then a column of random numbers in the range 0 to 1 could be generated and the sampling frame could be sorted by this column of numbers. Choose the units appearing in the first $n$ rows as the sample.
- **Tables of random numbers**. A table of random numbers could be used instead of a calculator to supply $n$ distinct random numbers in the range 1 to $N$.

Simple random sampling involves drawing units from the population one at a time without replacement. The number of different possible random samples of size $n$ is $\binom{N}{n}$ and each of these is equally likely.

## EXERCISE 3.2

Photocopies of the *Circles* worksheet (see photocopiable page 141) are required for this exercise. The worksheet shows a population of 60 circles of various sizes. The purpose of this exercise is to estimate the mean diameter of these 60 circles by measuring the diameters of a sample of five circles. Measuring the circles is easy – they have all been drawn with diameters that are some multiple of 0.5 cm. Two methods of selecting the sample of five circles will be compared.

**1** *Method 1: Non-random sampling*

  **a** Working independently, put a check mark ✓ in each of the five circles which in your judgement are representative of this population of circles.

  **b** Measure the diameters of the five circles you have selected and calculate the sample mean.

c Use a dotplot to display the sample means for the whole class.
(About 20 sample means should be adequate. If you are working alone, you could ask 20 people to select five circles for you and these can be measured later.)

2 *Method 2: Simple random sampling*
  a Number the circles from 1 to 60 in any order.
  b Produce five distinct random numbers in the range from 1 to 60.
  c Select the circles which have these random numbers, measure their diameter and calculate the sample mean.
  d Use a dotplot to display the sample means for the whole class.

3 Compare the dotplots drawn for questions 1 and 2.

4 Calculate the probability of obtaining any particular simple random sample of size 5 from this population of circles.



## The distribution of sample means

Within the population in Exercise 3.2, the circles have different sizes. This is an example of what is referred to as **natural variation** in the sizes of circles in the population. Individuals in a population are different from one another. For example, adult women (or adult men) differ from one another in height and weight; bolts manufactured in the same factory under the same conditions vary slightly in their diameter; and so on. This natural variation means that two samples drawn by the same sampling method from the same population will give somewhat different estimates of the population parameters. Whichever method was used to select the sample in Exercise 3.2, different samples may have different circles included, and the value of the sample mean will vary depending on the sizes of the particular circles included in the sample. This is known as **sampling variation**.

Both these sources of variability result in estimates of population parameters that are surrounded by some uncertainty. The more variability there is in the population, i.e. the greater the natural variation, the greater the uncertainty that surrounds our estimates. A method of sampling which produces estimates which differ from the true value in some systematic way is said to be **biased**. When circles are chosen in a non-random way (Method 1) we tend to ignore the fact that there are very many more small circles and this **selection bias** results in estimates which are on average too large. Although simple random sampling does not eliminate sampling variation, it does produce estimates that are centred around the true value and are on average equal to the unknown population parameter. Simple random sampling is an unbiased sampling method.

Suppose that $X_1, X_2, \ldots, X_n$ $(n \geq 2)$ are independent random variables from the same distribution. This represents the situation when we draw a simple random sample of size $n$ from a large population, where the random process used in selecting the sample allows us to assume independence. We say that these random variables are

**independent and identically distributed (iid).** In particular, suppose that each $X_i$ has expected value $\mu$ and variance $\sigma^2$.

Thus

$$E(X_i) = \mu \quad \text{and} \quad V(X_i) = \sigma^2 \quad \text{for } i = 1, 2, \ldots, n$$

Let

$$\overline{X} = \frac{1}{n}(X_1 + X_2 + \cdots + X_n)$$

The range space of the random variable $\overline{X}$ consists of all the possible values taken by the sample mean.

$$E(\overline{X}) = E\left(\frac{1}{n}(X_1 + X_2 + \cdots + X_n)\right) = \frac{1}{n}E(X_1) + \frac{1}{n}E(X_2) + \cdots + \frac{1}{n}E(X_n)$$

$$= \frac{1}{n}\mu + \frac{1}{n}\mu + \cdots + \frac{1}{n}\mu = \mu$$

and

$$V(\overline{X}) = V\left(\frac{1}{n}(X_1 + X_2 + \cdots + X_n)\right) = \frac{1}{n^2}V(X_1) + \frac{1}{n^2}V(X_2) + \cdots + \frac{1}{n^2}V(X_n)$$

$$= \frac{1}{n^2}\sigma^2 + \frac{1}{n^2}\sigma^2 + \cdots + \frac{1}{n^2}\sigma^2 = \frac{\sigma^2}{n}$$

These results are true for independent and identically distributed random variables from any probability distribution. They show how sampling variation can be described.

- The expected value of sample means calculated from simple random samples is equal to the population mean. In other words, if we were to repeatedly take simple random samples and use the observations in each sample to calculate a sample mean $\overline{x}$ to estimate $\mu$, then all these estimates would belong to a **sampling distribution** with mean $\mu$. Because the expected value of the sample mean (from simple random samples) is equal to the population parameter we say that the sample mean is an **unbiased estimator** of $\mu$. Appendix 5 (page 145) contains a proof that the sample variance is an unbiased estimator of $\sigma^2$.
- The variability of the sample means is less than the natural variation in the population:

$$\frac{\sigma^2}{n} < \sigma^2 \quad \text{so} \quad V(\overline{X}) < V(X_i)$$

The distribution of $X_i$

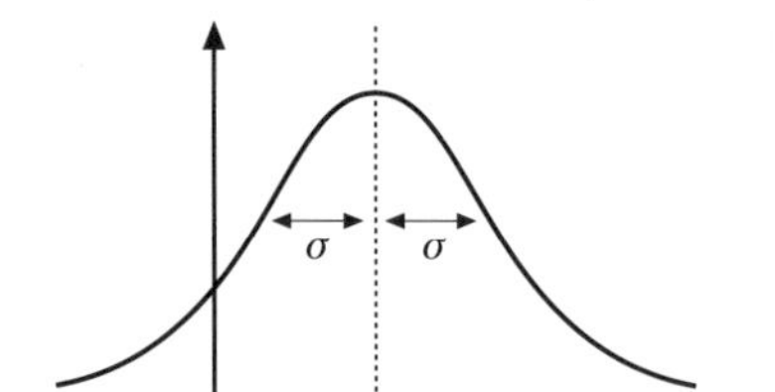


The sampling distribution of $\overline{X}$

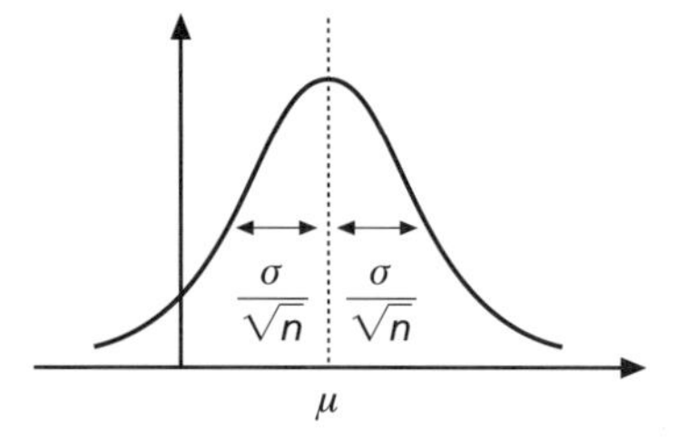


The variance of the sampling distribution varies directly as the amount of natural variation in the population, i.e. $V(\overline{X}) \propto \sigma^2$, and varies inversely as the sample size, i.e. $V(\overline{X}) \propto \frac{1}{n}$.

The square root of the variance of the sample means is often called the **standard error** of the mean:

$$\text{standard error} = \sqrt{V(\overline{X})} = \frac{\sigma}{\sqrt{n}}$$

The standard error is a numerical indicator of the precision of an estimate. A small standard error indicates that the estimate (sample mean $\overline{x}$) from any simple random sample of size $n$ is likely to be quite close to the true value of the parameter (population mean $\mu$), i.e. there is high precision. On the other hand, a large standard error indicates that the estimate from any particular sample could be farther away from the true value of the parameter, i.e. there is low precision.

So far we have not specified any particular probability distribution for the random variables $X_i$. However, if the random variables $X_i$ follow a Normal distribution then so does the random variable $\overline{X}$

$$\text{If } X_i \sim \mathrm{N}(\mu, \sigma^2) \quad \text{then} \quad \overline{X} \sim \mathrm{N}\left(\mu, \frac{\sigma^2}{n}\right)$$

*Example* A manufacturer produces ball bearings whose diameters are Normally distributed with a mean of 10.0 mm and standard deviation 0.1 mm.

**a** Calculate the probability that the diameter of a randomly selected ball bearing is between 9.9 mm and 10.1 mm.

**b** Calculate the standard error of the mean diameter of a random sample of four ball bearings.

**c** What is the probability that the mean diameter of a random sample of four ball bearings is between 9.9 mm and 10.1 mm?

*Solution*

**a** Let $X$ be the random variable representing the diameter of the ball bearings.

$$\mathrm{P}(9.9 < X < 10.1) = \mathrm{P}\left(\frac{9.9 - 10.0}{0.1} < Z < \frac{10.1 - 10.00}{0.1}\right)$$
$$= \Phi(1.0) - \Phi(-1.0) = 2(\Phi(1.0) - 0.5) = 2(0.8413 - 0.5)$$
$$= 2 \times 0.3413 = 0.6826$$

**b** Standard error of the mean diameter $= \dfrac{\sigma}{\sqrt{n}} = \dfrac{0.1}{\sqrt{4}} = 0.05$ mm

**c**

$$\mathrm{P}(9.9 < \overline{X} < 10.1) = \mathrm{P}\left(\frac{9.9 - 10.0}{0.05} < Z < \frac{10.1 - 10.0}{0.05}\right)$$
$$= \Phi(2.0) - \Phi(-2.0) = 2(\Phi(2.0) - 0.5) = 2(0.9772 - 0.5)$$
$$= 2 \times 0.4772 = 0.9544$$

## EXERCISE 3.3

1 $X \sim \mathrm{N}(10, 9)$. A computer program generated a simple random sample of four observations from this distribution:

8.1814 11.9008 7.5648 10.4211

**a** Calculate the sample mean (answer correct to two decimal places).

**b** Calculate the standard error of the mean.

**c** Calculate the probability that the mean of a simple random sample of four observations from this distribution would take a value less than that calculated in **a**.

**2** The heights of men in a certain population are Normally distributed with a mean of 175 cm and a standard deviation of 7.5 cm.

**a** If a man is chosen at random, what is the probability that his height is greater than 180 cm?

If a simple random sample of nine men is drawn from this population,

**b** calculate the standard error of the mean height,

**c** calculate the probability that the sample mean height exceeds 180 cm.

**3** A certain brand of light bulb has lifetimes which are Normally distributed with a mean of 1000 hours and standard deviation 100 hours.

**a** Calculate the probability that a randomly selected light bulb has a lifetime of less than 900 hours.

**b** Calculate the standard error of the mean lifetime of a random sample of six light bulbs.

**c** Calculate the probability that the mean lifetime of a random sample of six light bulbs is less than 900 hours.

**4** $X \sim \mathrm{N}(\mu, \sigma^2)$.

**a** Calculate $\mathrm{P}(\mu - 0.1\sigma < X < \mu + 0.1\sigma)$.

A random sample of size $n$ is drawn from this population.

**b** If $n = 100$, state the distribution of $\overline{X}$ and calculate $\mathrm{P}(\mu - 0.1\sigma < \overline{X} < \mu + 0.1\sigma)$.

**c** Calculate the sample size $n$ if we require $\mathrm{P}(\mu - 0.1\sigma < \overline{X} < \mu + 0.1\sigma) \geq 0.95$.

## *The Central Limit Theorem*

We now state a truly remarkable result:

> Let $X_1, X_2, \ldots, X_n$ be a sequence of independent and identically distributed random variables, each with expected value $\mu$ and variance $\sigma^2$. Then, for sufficiently large $n$,
>
> $$\overline{X} \sim \mathrm{N}\left(\mu, \frac{\sigma^2}{n}\right) \text{ approximately}$$
>
> and, the larger $n$ is, the better the approximation will be.

What this means is that, regardless of the parent distribution of the original random variables, the sampling distribution of $\overline{X}$ is approximately Normally distributed provided $n$ is large enough. This is true even when the $X_i$s are discrete random variables as was the case, for example, when the Normal approximation to the Binomial was demonstrated in Chapter 2. The shape of the parent distribution is very important in determining how large $n$ should be. In most cases a sample size of around 20 or 30 values will give a good approximation. However, if the parent distribution is highly skewed then a larger sample size will be required for a good approximation.

Of course, if the parent distribution is Normally distributed to begin with, then we do not need to make use of the Central Limit Theorem because, as we stated earlier, in such cases $\overline{X} \sim N\left(\mu, \dfrac{\sigma^2}{n}\right)$ exactly. The Central Limit Theorem is sometimes stated in an alternative way which is equivalent to the above, namely: for sufficiently large $n$

$$\sum_{i=1}^{n} X_1 \sim N(n\mu, n\sigma^2) \text{ approximately}$$

*Example* The discrete random variable $X_i$ represents the score on the $i$th throw of a fair die.

**a** Illustrate the theoretical distribution of $X_i$ with a bar diagram.

**b** Write down the probability $P(X_i < 3)$.

**c** Use the Central Limit Theorem (CLT) to calculate the approximate probability that the total score of 30 throws of the die is less than 90.

**d** Calculate the approximate probability that the mean of 30 throws of the die is less than 3.

*Solution*

**a**

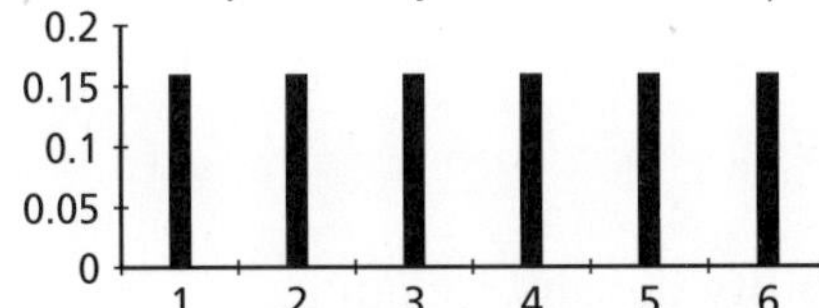


This is a Uniform distribution.

**b** $P(X_i < 3) = \frac{1}{3}$

**c** $$E(X_i) = 1\times\frac{1}{6} + 2\times\frac{1}{6} + 3\times\frac{1}{6} + 4\times\frac{1}{6} + 5\times\frac{1}{6} + 6\times\frac{1}{6} = \frac{7}{2}$$

$$V(X_i) = 1^2\times\frac{1}{6} + 2^2\times\frac{1}{6} + 3^2\times\frac{1}{6} + 4^2\times\frac{1}{6} + 5^2\times\frac{1}{6} + 6^2\times\frac{1}{6} - \left(\frac{7}{2}\right)^2 = \frac{35}{12}$$

Let $T = \Sigma_{i=1}^{30} X_i$,

$$E(T) = E\left(\sum_{i=1}^{30} X_i\right) = \sum_{i=1}^{30} E(X_i) = 30 \times \frac{7}{2} = 105$$

$$V(T) = V\left(\sum_{i=1}^{30} X_i\right) = \sum_{i=1}^{30} V(X_i) = 30 \times \frac{35}{12} = 87.5$$

By CLT, $T \sim N(105, 87.5)$ approximately.

$$P(T < 90) = P\left(Z < \frac{89.5 - 105}{\sqrt{87.5}}\right) \quad \text{Note the continuity correction.}$$

$$= \Phi(-1.66) = 1 - \Phi(1.66) = 1 - 0.9515 = 0.0485$$

Note that, without the continuity correction, $P(T < 90) = \Phi(-1.60) = 0.0548$.

**d** $\overline{X} = \frac{1}{30}T$

$$E(\overline{X}) = \frac{1}{30}E(T) = \frac{1}{30} \times 105 = 3.5$$

$$V(\overline{X}) = V(\tfrac{1}{30}T) = \frac{1}{900}V(T) = \frac{1}{900} \times 87.5 = \frac{7}{72} \approx 0.09722$$

By CLT, $\overline{X} \sim N(3.5, 0.09722)$ approximately.

Whereas the range space of $T$ consists of whole numbers {30, 31, 32, …, 179, 180} the range space of $\overline{X}$ consists of the numbers $\left\{\frac{30}{30}, \frac{31}{30}, \frac{32}{30}, \ldots, \frac{179}{30}, \frac{180}{30}\right\}$ and, although this is discrete, care should be taken when applying continuity corrections. Strictly, in this case, $P(\overline{X} < 3)$ means $P\left(\overline{X} < \frac{90}{30}\right)$ or $P\left(\overline{X} \le \frac{89}{30}\right)$ so applying a continuity correction would mean calculating

$$P(\overline{X} < 3) = P\left(Z < \frac{\left(\frac{90}{30} - \frac{1}{60}\right) - 3.5}{\sqrt{0.09722}}\right) = \Phi(-1.66) = 0.0485$$

Without this 'continuity correction' we get $P(\overline{X} < 3) = \Phi(-1.60) = 0.0548$. Note that these are the same answers as in part **c**.

*Warning*

The ±0.5 continuity correction should only be used when the range space of a discrete random variable consists of integers.

## EXERCISE 3.4

**1** A continuous random variable $X$ has the probability density function:

$$f(x) = \begin{cases} \frac{1}{5} & 0 \le x \le 5 \\ 0 & \text{otherwise} \end{cases}$$

**a** Sketch $f(x)$.
**b** Calculate the mean and variance of $X$.

Random samples of 25 observations are taken of this population.

**c** State the approximate sampling distribution of $\overline{X}$.
**d** Calculate (**i**) $P(X < 2)$ (**ii**) $P(\overline{X} < 2)$.

**2** The graph shows the probability density function for a continuous random variable $X$ representing the lifetime in hours of a certain component.

$$\mu = E(X) = 10 \text{ and } \sigma^2 = V(X) = 100$$

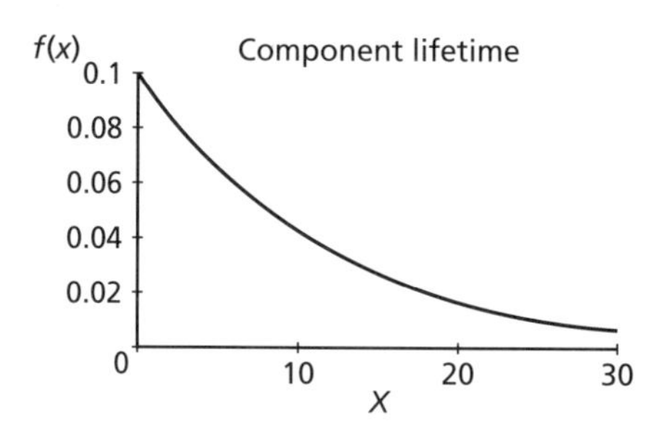


**a** State the approximate distribution of the mean lifetime $\overline{X}$ for random samples of 50 such components.
**b** Calculate (**i**) $P(\overline{X} < 8)$ (**ii**) $P(\overline{X} > 11)$.

**3** The discrete random variables $X_i (i = 1, 2, \ldots, 30)$ are iid $X_i \sim \text{Bin}(20, 0.5)$.

**a** Write down $E(X_i)$ and $V(X_i)$.
**b** State the approximate distribution of $T = \Sigma_{i=1}^{30} X_i$.
**c** Calculate $P(T \le 300)$.
**d** Hence, state $P(\overline{X} \le 10)$ where $\overline{X} = \frac{1}{30}T$.

**4** The time $X$ seconds between consecutive vehicles passing a fixed point on a motorway is thought to be a random variable with probability density function

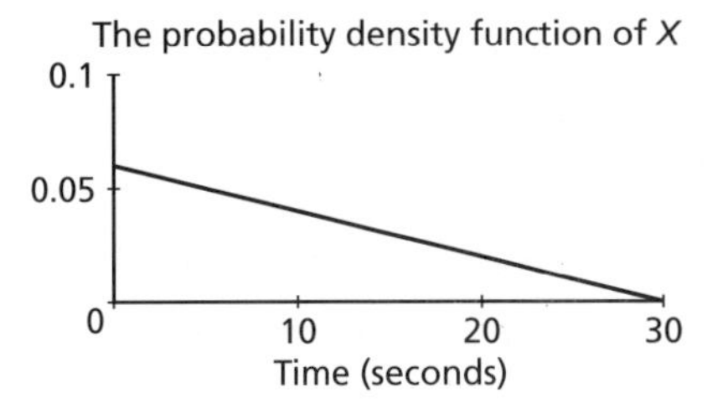


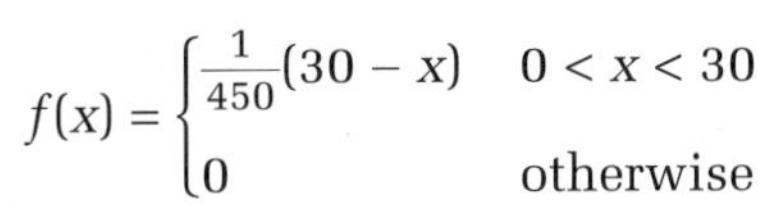

$$f(x) = \begin{cases} \frac{1}{450}(30 - x) & 0 < x < 30 \\ 0 & \text{otherwise} \end{cases}$$

**a** Calculate the mean and variance of $X$.

Suppose random samples of 25 such time intervals are taken.

**b** Determine the approximate sampling distribution of $\overline{X}$.

**c** Why is it essential to insist on *random* samples when using the Central Limit Theorem?

**d** Calculate **(i)** $P(X < 7)$ **(ii)** $P(\overline{X} < 7)$.

**5** A continuous random variable $X$ has the probability density function:

$$f(x) = \begin{cases} 3x^2 & 0 \leq x \leq 1 \\ 0 & \text{otherwise} \end{cases}$$

**a** Sketch the pdf.

**b** Determine the mean and variance of $X$.

Suppose random samples of 36 observations of this random variable are taken.

**c** Determine the approximate sampling distribution of $\overline{X}$.

**d** Why is it essential to insist on *random* samples when using the Central Limit Theorem?

**e** Calculate **(i)** $P(X < 0.6)$ **(ii)** $P(\overline{X} < 0.6)$.

**6** The wages paid to production workers in a large factory have a mean of £175 per week and a standard deviation of £20 per week.

**a** State the approximate distribution for the mean weekly wage of random samples of 30 workers.

**b** Calculate the approximate probability that the mean weekly wage of a random sample of 30 such workers is more than £180.

**7** A discrete random variable has the Bernoulli distribution $X \sim \text{Ber}(0.4)$ when

| $x$ | 0 | 1 |
|---|---|---|
| $P(X = x)$ | 0.6 | 0.4 |

**a** Calculate the mean and variance for this distribution.

Random samples of 25 observations $X_i$ ($i = 1, 2, \ldots, 25$) are taken where $X_i \sim \text{Ber}(0.4)$.

**b** Determine the approximate sampling distribution of $T = \Sigma_{i=1}^{25} X_i$ and calculate $P(T \leq 15)$.

The random variable $W \sim \text{Bin}(25, 0.4)$.

**c** Use the Normal approximation to the Binomial and calculate $P(W \leq 15)$. Compare your answer with that for **b**.

## *A confidence interval for a population mean*

Usually we have just one sample from a population, which we must use to estimate the unknown population mean. Although the sample mean gives an unbiased **point estimate** of the population mean, this is not completely satisfactory because the sample mean and the population mean will not generally be equal.

Instead of relying on one number, we might try to use the sample data to identify a range of plausible values for the population mean. Such a range of values is called an **interval estimate** or a **confidence interval**. We would want to be sure that, whenever we calculated a confidence interval, it was very likely to contain the true value of the population mean.

The Central Limit Theorem (CLT) says that, for large enough $n$, when $X_1, X_2, \ldots, X_n$ are independent and identically distributed random variables, each with expected value $\mu$ and variance $\sigma^2$, then

$$\overline{X} \sim \mathrm{N}\left(\mu, \frac{\sigma^2}{n}\right) \text{ approximately}$$

This means that

$$\frac{\overline{X} - \mu}{\sigma/\sqrt{n}} \sim \mathrm{N}(0, 1) \text{ approximately}$$

Note that if the sample comes from a population that is Normally distributed we shall not need to make use of the CLT because in such cases the sampling distribution of $\overline{X}$ is known exactly. The power of the CLT is that we do not need to know whether or not the parent population is Normally distributed. All we need to assume is that each member of the sample is independently and identically distributed as long as $n$ is large enough. Independence may be assumed if the sample selected is a *random* sample.

Although, in practice, the population standard deviation $\sigma$ is usually unknown, we shall assume for the moment that we know its value. This means that the standard error $\sigma/\sqrt{n}$ is known exactly and does not have to be estimated from the sample data. (The construction of confidence intervals when $\sigma$ is unknown, thus requiring the standard error to be estimated, is discussed in the Advanced Higher Unit, Statistics 2.)

From Table 3 (Appendix 1, page 139)

$$\mathrm{P}\left(-1.96 < \frac{\overline{X} - \mu}{\sigma/\sqrt{n}} < 1.96\right) = 0.95$$

$$\mathrm{P}\left(-1.96\frac{\sigma}{\sqrt{n}} < \overline{X} - \mu < 1.96\frac{\sigma}{\sqrt{n}}\right) = 0.95$$

$$\mathrm{P}\left(-1.96\frac{\sigma}{\sqrt{n}} < \mu - \overline{X} < 1.96\frac{\sigma}{\sqrt{n}}\right) = 0.95$$

$$\mathrm{P}\left(\overline{X} - 1.96\frac{\sigma}{\sqrt{n}} < \mu < \overline{X} + 1.96\frac{\sigma}{\sqrt{n}}\right) = 0.95$$

In other words, there is probability 0.95 that the interval $\left(\overline{X} - 1.96\frac{\sigma}{\sqrt{n}}, \overline{X} + 1.96\frac{\sigma}{\sqrt{n}}\right)$ contains the true value of the population mean $\mu$. This is a **95% confidence interval** for the population mean. A 95% confidence interval will not always contain the true value of the population parameter. In fact, averaged over our lifetime, only about 95% of the intervals we calculate will capture the true value.

The degree of confidence and the width of the confidence interval can be varied as follows:

Confidence interval for population mean is $\bar{X} \pm k\frac{\sigma}{\sqrt{n}}$.

For 90% confidence interval use $k = 1.64$.
For 95% confidence interval use $k = 1.96$.
For 99% confidence interval use $k = 2.58$.

The required value of $k$ can also be found in Table 4 (Appendix 1, page 140). For a $100(1 - 2p)\%$ confidence interval use $k = z_p$.

*Example 1* The heights of men in a certain population are thought to be Normally distributed with a standard deviation of 10 cm. The heights (in cm) of a random sample of five men are:

159 172 177 170 162

**a** Calculate **(i)** the sample mean, **(ii)** the standard error of the mean.

**b** Construct a 95% confidence interval (CI) for the mean height of men in this population.

*Solution*

**a** (i) $\bar{x} = \frac{\Sigma x}{n} = \frac{840}{5} = 168$ cm (ii) Standard error $= \frac{\sigma}{\sqrt{n}} = \frac{10}{\sqrt{5}} = 4.47$

**b** The heights $X_i \sim N(\mu, \sigma^2)$ so $\bar{X} \sim N\left(\mu, \frac{\sigma^2}{n}\right)$ exactly

$$95\% \text{ CI is } \left(\bar{X} - 1.96\frac{\sigma}{\sqrt{n}}, \bar{X} + 1.96\frac{\sigma}{\sqrt{n}}\right) = (168 - 1.96 \times 4.47, 168 + 1.96 \times 4.47)$$
$$= (159.24, 176.76)$$

95% of such CIs capture the population mean height. It is very likely that the mean height of men in this population lies between 159.24 and 176.76 cm.

*Example 2* The daily yield of a chemical manufactured at a certain chemical plant is recorded for 50 days. The sample mean is 871 tonnes and the population standard deviation is known to be 21 tonnes. Construct a 99% confidence interval (CI) for the mean daily yield, $\mu$ tonnes.

*Solution*

The exact form of the distribution of yields $X$ is not known.

However, by the Central Limit Theorem, $\bar{X} \sim N\left(\mu, \frac{\sigma^2}{n}\right)$ approximately and we assume the approximation is reasonably good since $n = 50$ is moderately large.

The 99% CI is

$$\left(\bar{X} - 2.58\frac{\sigma}{\sqrt{n}}, \bar{X} + 2.58\frac{\sigma}{\sqrt{n}}\right) = \left(871 - 2.58\frac{21}{\sqrt{50}}, 871 + 2.58\frac{21}{\sqrt{50}}\right)$$
$$= (863.3, 878.7)$$

99% of such CIs capture the population mean daily yield so we can say with 99% confidence that the true mean daily yield lies between 863.3 and 878.7 tonnes.

## EXERCISE 3.5A

**1** The heights of women in a certain population are assumed to be Normally distributed with a standard deviation of 7.5 cm. The heights (in cm) of a random sample of five women are:

157 167 169 163 151

**a** Calculate **(i)** the sample mean, **(ii)** the standard error of the mean.

**b** Construct a 95% confidence interval for the mean height of women in this population, and interpret your interval clearly.

**2** A random sample of eight leaves was collected from a laurel hedge and the length of each leaf measured to the nearest millimetre.

70 93 90 102 98 114 126 120

Assume the lengths of this type of leaf are Normally distributed with standard deviation 20 mm.

**a** Calculate **(i)** the sample mean, **(ii)** the standard error.

**b** Construct a 90% confidence interval for the mean length of this type of laurel leaf.

**c** Construct a 95% confidence interval for the mean and compare your answer with part **b**.

**3** A random sample of ten packets of crisps of a certain brand is taken and the contents of each packet are weighed. The contents (in grams) are as follows:

25.88 25.51 26.01 25.25 25.40 26.15 25.70 25.82 26.45 25.63

Construct a 95% confidence interval for the true mean mass of crisps in packets of this brand, and interpret this confidence interval clearly.
Assume that the masses are Normally distributed with a known standard deviation of 0.35 g.

**4** A random sample of nine male students was selected and their weights were recorded (in kg):

59.1 59.0 68.7 61.8 80.7 53.9 83.1 56.4 75.5

Assuming the weights of male students in this population are Normally distributed with standard deviation 10 kg, construct a 95% confidence interval for the population mean weight. Interpret this confidence interval carefully.

**5** A group of fifteen male volunteers had their body potassium measured in a whole-body monitor as part of a study to investigate body composition. The measurements (in Becquerel per kg body weight) are shown here:

| | | | | |
|---|---|---|---|---|
| 133.5 | 149.6 | 120.9 | 126.8 | 138.1 |
| 189.0 | 184.7 | 109.8 | 171.6 | 148.7 |
| 157.0 | 146.6 | 150.1 | 143.0 | 138.8 |

**a** Draw a boxplot to display these data.

**b** Assuming that this group can be regarded as a random sample and assuming potassium levels in men are distributed $N(\mu, 20^2)$, obtain a 95% confidence interval for $\mu$. Interpret this confidence interval carefully.

## EXERCISE 3.5B

**1** As part of a large study of how environmental pollutants affect animals, the thickness of 65 pelican egg shells was measured and the sample mean was 0.32 mm. From previous studies it is known that the standard deviation is 0.08 mm.

**a** State the approximate sampling distribution of mean shell thickness.

**b** Construct a 95% confidence interval for the mean shell thickness. Interpret this confidence interval carefully.

**2** A basic skills maths test has been designed to produce scores on a scale from 0 to 100 with standard deviation 20. The scores for a random sample of pupils taking the test were:

| | | | | | | | | | | | |
|---|---|---|---|---|---|---|---|---|---|---|---|
| 80 | 59 | 65 | 47 | 79 | 53 | 63 | 49 | 88 | 31 | 55 | 58 |
| 51 | 30 | 33 | 72 | 40 | 61 | 45 | 93 | 85 | 69 | 74 | 66 |

**a** Draw a stem and leaf diagram for these data and comment on its shape.

**b** Calculate the sample mean for this group of pupils.

**c** State the approximate sampling distribution for the mean score.

**d** Construct **(i)** 90%, **(ii)** 95% and **(iii)** 99% confidence intervals for the true mean score of pupils taking the test and compare them.

**3** Past experience has indicated that the breaking strength of a fibre has standard deviation 3 units. A random sample of 25 fibres was tested and the sample mean breaking strength was 148 units.

**a** Construct a 95% confidence interval for the mean breaking strength of this type of fibre.

**b** It is planned to use these fibres in an application which requires fibres with a minimum breaking strength of 150 units. Should they be used? Give a reason for your answer.

**4** The weights (in kilograms) of a random sample of ten castings produced in an iron foundry were:

| | | | | | | | | | |
|---|---|---|---|---|---|---|---|---|---|
| 19.8 | 20.3 | 20.6 | 21.1 | 19.3 | 19.6 | 20.1 | 20.8 | 21.1 | 21.3 |

From past experience it is known that such castings have a standard deviation of 0.6 kg. The foundry claims that its castings have a mean weight of 20.0 kg.

**a** Illustrate these data with a boxplot and indicate 20.0 kg clearly on your diagram.

**b** Construct a 99% confidence interval for the true mean weight of castings.

**c** Is the foundry's claim justified? Give a reason for your answer.

**5** Radiocarbon, a naturally occurring radioactive element, has a global environmental level of 257 becquerel per kilogram of carbon. As part of a study to investigate the dispersal of artificially produced radiocarbon around a nuclear fuel reprocessing plant, 14 environmental samples were collected and their levels of radiocarbon measured. The results are given below:

| | | | | | | |
|---|---|---|---|---|---|---|
| 286.2 | 284.9 | 282.3 | 279.6 | 280.2 | 278.3 | 278.7 |
| 277.5 | 279.1 | 282.1 | 284.5 | 283.7 | 280.0 | 276.9 |

**a** Illustrate these data with a boxplot and show the global level of 257 Bq/kg C on the diagram.

**b** Construct a 95% confidence interval for the true mean level of radiocarbon surrounding the reprocessing plant if the standard deviation is known to be 3 Bq/kg C.

**c** Is the level of radiocarbon around the plant higher than the global environmental level? Give a reason for your answer.

6 As part of a study to produce diagnostic criteria for patients with a particular heart problem, a sample of 25 healthy, 30–39-year-old males had electrocardiograms (ECGs) taken. One important component of the resulting ECG is the major amplitude of the signal. The major amplitudes recorded in this sample were as follows:

| | | | | | | | | | | | | |
|---|---|---|---|---|---|---|---|---|---|---|---|---|
| 2.8 | 1.4 | 2.0 | 2.5 | 2.1 | 1.4 | 2.4 | 1.8 | 1.3 | 2.4 | 1.9 | 1.5 | 1.2 |
| 1.8 | 2.3 | 0.8 | 1.7 | 2.1 | 2.0 | 2.1 | 1.4 | 1.9 | 2.3 | 2.9 | 1.0 | |

It is known from past studies that such amplitudes have a standard deviation of 0.5.

**a** Illustrate these data with a stem and leaf diagram and comment on its shape.

**b** Construct a 95% confidence interval of the population mean value of the major amplitude. Interpret this confidence interval clearly.

7 In an experiment on feeding pregnant beef cows with molassed sugar-beet pulp, the weight gains (in kilograms) of twelve cows were as follows (a negative value means the cow lost weight):

| | | | | | |
|---|---|---|---|---|---|
| −0.667 | 1.048 | 0.286 | 0.667 | −0.143 | 0.619 |
| 0.333 | 0.429 | 1.238 | 4.290 | 0.524 | 0.952 |

**a** Construct a 95% confidence interval for the mean weight gain that would be experienced by similar cows on this diet. You may assume that the population standard deviation of weight gained is 0.52 kg.

**b** Illustrate the data with a boxplot, showing any possible outliers on your diagram.

**c** It should be a matter of routine to display data before calculating confidence intervals. Re-calculate the 95% confidence interval omitting the outlier, and comment.

8 A Normally distributed random variable has an unknown mean $\mu$ and a known variance $\sigma^2 = 16$. Calculate the sample size required to construct a 95% confidence interval for the population mean, which has a total width of 2.0.

## A confidence interval for a population proportion

When working with large samples, the Central Limit Theorem provides a reliable method for constructing approximate confidence intervals for a population proportion.

Suppose $X$ is the total number of 'successes' in $n$ independent trials, each with success probability $p$. A suitable model is $X \sim \text{Bin}(n, p)$.

This represents the situation when, for example, a random sample of $n$ voters is asked:

'Do you intend to vote Conservative at the next Election?'

where $p$ is the true proportion of the electorate that support this party.

Let $X_i$ be the number of successes in the $i$th trial ($i = 1, 2, \ldots, n$). So $X_i = 1$ if the voter answers 'Yes' and $X_i = 0$ if the voter answers 'No'. Now

$$P(X_i = 1) = p \quad \text{and} \quad P(X_i = 0) = 1 - p = q$$
$$E(X_i) = 1 \times p + 0 \times (1 - p) = p$$

and

$$V(X_i) = 1^2 \times p + 0^2 \times (1 - p) - p^2 = p(1 - p) = pq$$

Now $X = \sum_{i=1}^{n} X_i$ and $X_1, X_2, \ldots, X_n$ are independent random variables, where

$$\mathrm{E}(X) = \mathrm{E}\left(\sum_{i=1}^{n} X_i\right) = \sum_{i=1}^{n} \mathrm{E}(X_i) = np \quad \text{and} \quad \mathrm{V}(X) = \mathrm{V}\left(\sum_{i=1}^{n} X_i\right) = \sum_{i=1}^{n} \mathrm{V}(X_i) = npq$$

So, by applying the Central Limit Theorem (CLT), we can justify:

for large $n$, $X \sim \mathrm{N}(np, npq)$ approximately

The approximation is satisfactory if $n \geq 20$, $np \geq 5$ and $nq \geq 5$. This is a justification of the Normal approximation to the Binomial, which was introduced in Chapter 2.

*Example 1* A random sample of 100 voters was asked whether they supported a certain political party. If the true proportion of voters who support this party is 0.35, calculate the probability that fewer than 30 voters in the sample indicated their support for the party.

*Solution*

Let $X$ represent the number of supporters in the sample. $X \sim \mathrm{Bin}(100, 0.35)$.
Since $n$ is large and $np = 35$ and $nq = 65$ we can use $X \sim \mathrm{N}(35, 22.75)$ approximately.

$$\mathrm{P}(X < 30) = \mathrm{P}\left(Z < \frac{29.5 - 35}{\sqrt{22.75}}\right) \quad \text{Note the continuity correction.}$$

$$= 1 - \mathrm{P}(Z < 1.15) = 0.1251$$

Usually, however, the true population proportion is unknown and our task is to estimate it using the information in a sample.

The obvious estimator for $p$ from a sample is the sample proportion of 'successes' $\frac{X}{n}$.

$$\mathrm{E}\left(\frac{X}{n}\right) = \frac{1}{n}\mathrm{E}(X) = \frac{np}{n} = p \quad \text{and} \quad \mathrm{V}\left(\frac{X}{n}\right) = \frac{1}{n^2}\mathrm{V}(X) = \frac{npq}{n^2} = \frac{pq}{n}$$

So, for large enough $n$,

$$\frac{X}{n} \sim \mathrm{N}\left(p, \frac{pq}{n}\right) \text{ approximately} \quad \text{and} \quad \frac{\frac{X}{n} - p}{\sqrt{\frac{pq}{n}}} \sim \mathrm{N}(0, 1) \text{ approximately}$$

The quantity $\sqrt{\frac{pq}{n}}$ is known as the **standard error** of the sample proportion $\frac{X}{n}$.

From a sample, calculate an estimate $\hat{p} = \frac{x}{n}$ and replace $p$ with this estimate.
Also $\hat{q} = 1 - \hat{p}$. An approximate 95% confidence interval for $p$ is given by

$$\left(\hat{p} - 1.96\sqrt{\frac{\hat{p}\hat{q}}{n}}, \hat{p} + 1.96\sqrt{\frac{\hat{p}\hat{q}}{n}}\right)$$

provided $n \geq 20$, $np \geq 5$, and $nq \geq 5$. The quantity $\sqrt{\frac{\hat{p}\hat{q}}{n}}$ is known as the **estimated standard error**.

*Example 2* A study was carried out of an adhesive commonly used to affix orthodontic brackets. Of the 418 orthodontic brackets affixed using this adhesive, 39 failed within six months. Construct a 95% confidence interval for the probability of a failure within six months.

*Solution*

$$\hat{p} = \frac{39}{418} = 0.0933 \qquad \hat{q} = 1 - 0.0933 = 0.9067$$

$$\text{Estimated standard error} = \sqrt{\frac{0.0933 \times 0.9067}{418}} = 0.0142$$

95% confidence interval is $(0.0933 - 1.96 \times 0.0142,\ 0.0933 + 1.96 \times 0.0142)$
$= (0.065,\ 0.121)$

95% of such CIs capture the population proportion. So, when this adhesive is used, it is very likely that the proportion which fail within six months lies between 0.065 and 0.121.

Note that when constructing confidence intervals by this method it is common practice not to involve any correction for continuity. For large samples, this is satisfactory.

## EXERCISE 3.6B

**1** A coin is tossed 100 times and Heads appears 64 times.

**a** Calculate
 **(i)** the sample proportion,
 **(ii)** the estimated standard error.

**b** Construct a 95% confidence interval for the true proportion of Heads.

**c** Do you think this is a fair coin? Give a reason for your answer.

**2** A random sample of 30 pupils at a large secondary school were asked how they had travelled to school that day. Nine replied they had travelled by car.

**a** Calculate
 **(i)** the sample proportion,
 **(ii)** the estimated standard error.

**b** Construct an approximate 95% confidence interval for the true proportion of pupils at this school who travel to school by car. Interpret this confidence interval clearly.

**3** A sample survey of 1000 Scottish voters found that 220 people said they intended to vote Liberal Democrat. Construct an approximate 90% confidence interval for the true proportion of Scottish voters who intend to vote Liberal Democrat. List some reasons why this confidence interval could give a biased measure of the actual behaviour of Scottish voters.

**4** As part of a market research exercise, a random sample of 500 households were asked 'Do you own an automatic dishwasher?' and 237 'Yes' replies were counted. Construct

**a** 90%, **b** 95%, and **c** 99%

confidence intervals for the true population proportion of households that have an automatic dishwasher. Compare your three answers.

**5** **a** A sports psychologist surveyed a random sample of 100 residents of a large housing estate. She assessed 45 of these individuals as 'physically active' according to strict criteria. Find a 95% confidence interval for the proportion of physically active people in the population.

**b** Of the 100 people surveyed, 61 were female and 39 were male. 24 females and 21 males were physically active. Obtain 95% confidence intervals for the population proportions of physically active females and males. Compare these confidence intervals. Do you think there is a difference between the proportion of females and males who are physically active? Give a reason for your answer.

**6** **a** Many people who undergo orthodontic treatment do so to improve their dental attractiveness (e.g. by straightening their front teeth). In a recent survey, a random sample of 303 new orthodontic patients were assessed according to a standard protocol. The consultant orthodontist rated 28 of them as dentally 'attractive' before treatment began. In the population of new orthodontic patients find an interval estimate for the proportion who are dentally 'attractive' before treatment begins.

**b** Unusually, in this survey, the patients were invited to rate their own dental attractiveness by the same criteria as used by the consultant. 255 of them over-rated their own dental 'attractiveness'. Construct a 95% confidence interval for the proportion of new orthodontic patients who would over-rate their own attractiveness, and comment on what you find.

**7** In a survey, 22 pairs of identical female twins, aged 40–49 years old, were asked about their political opinions. The twins in fourteen of the pairs had similar opinions, whilst the twins in the other eight pairs did not.

**a** Construct a 95% confidence interval for the population proportion of identical twins who hold similar political opinions.

**b** Do you think identical twins are equally likely to hold similar and dissimilar political opinions? Give a reason for your answer.

## *Stratified random sampling*

Consider a finite population consisting of the 25 numbers shown.

| | | | | |
|---|---|---|---|---|
| 34 | 24 | 63 | 57 | 45 |
| 44 | 23 | 54 | 25 | 53 |
| 65 | 26 | 67 | 53 | 63 |
| 34 | 43 | 34 | 36 | 27 |
| 47 | 35 | 47 | 56 | 65 |

In this example, the population mean can be easily calculated directly and is 44.8. In practice, we shall be dealing with much larger populations and in such situations we are more likely to estimate the population mean from a sample.

Notice that when this population of 25 numbers is rearranged its structure can be appreciated. There are five **strata** each consisting of five numbers, namely, the twenties, thirties, forties, fifties and sixties. The variability within each stratum is much less than in the population as a whole.

| | | | | |
|---|---|---|---|---|
| 23 | 24 | 25 | 26 | 27 |
| 34 | 34 | 34 | 35 | 36 |
| 43 | 44 | 45 | 47 | 47 |
| 53 | 53 | 54 | 56 | 57 |
| 63 | 63 | 65 | 65 | 67 |

Suppose our plan is to estimate the population mean from a simple random sample of five numbers drawn from the population. The sample mean calculated would lie somewhere in the interval

$$\frac{23 + 24 + 25 + 26 + 27}{5} \leq \bar{x} \leq \frac{63 + 63 + 65 + 65 + 67}{5}$$

$$25.0 \leq \bar{x} \leq 64.6$$

Another method of choosing a sample of five numbers is to select one number at random from each of the strata. The sample mean calculated as a result of this method of sampling would be in the interval

$$\frac{23 + 34 + 43 + 53 + 63}{5} \leq \bar{x} \leq \frac{27 + 36 + 47 + 57 + 67}{5}$$

$$43.2 \leq \bar{x} \leq 46.8$$

With this method of sampling we have made sure that each stratum in the population is fairly represented in the sample and as a result the estimates of the population mean lie in an interval which is narrower and which more closely surrounds the true value 44.8. Of course, to take advantage of the structure of the set of numbers, we require to know that there are different strata in the population and we require to know which particular numbers lie in which strata.

**Stratified random sampling** is used when the population can be divided into non-overlapping groups or strata, where the units in each stratum are thought to have similar characteristics. For example, political views tend to be very different for different age groups, so we could use age as a **stratification variable** with categories 18–29, 30–39, …, 60–69, 70 or over. Every member of the population belongs to one and only one of these strata. There will be less variability within these strata than in the population as a whole. Stratification can improve the precision of estimation.

A sampling frame is drawn up for each stratum separately and a simple random sample is then drawn from each stratum. Usually, a common sampling fraction is used in all the strata. This is known as **proportional allocation** because the proportion of the total sample that comes from a particular stratum is the same as the proportion of the total population that comes from that stratum. In order to use stratification in practice, we need to know (or suspect) that the population consists of strata which differ substantially in their opinions or behaviour. We also need to be able to determine the stratum into which each individual falls.

For example, before a particular university began a programme of upgrading its sports facilities, it was decided to conduct a sample survey to gather the opinions of students. The population consists of all the students enrolled at the university, the sampling units are individual students and a sampling frame is available from the university's matriculation records. It is thought that male students may have similar views and that female students may have opinions that, although common among females, are different from those of the male students. It is also thought that, to be representative, the sample should include a fair number of students from each Faculty. Records show the following numbers of full-time students:

| **Population** | Arts | Law & Finance | Science | Engineering | Medicine | Divinity |
|---|---|---|---|---|---|---|
| Male | 1251 | 470 | 2061 | 1107 | 719 | 40 |
| Female | 2639 | 684 | 2089 | 186 | 1067 | 46 |

A questionnaire will be mailed to 10% of students. A simple random sample is drawn from each of the twelve strata. The number of students randomly chosen from each stratum is:

| **Sample** | Arts | Law & Finance | Science | Engineering | Medicine | Divinity |
|---|---|---|---|---|---|---|
| Male | 125 | 47 | 206 | 111 | 72 | 4 |
| Female | 264 | 68 | 209 | 19 | 107 | 5 |

Like simple random sampling, stratified random sampling produces estimates which are unbiased. However, its main advantage is that it improves precision. Estimates will be, on average, closer to the true value of the population parameter.

*Warning*

When using stratified random sampling, do not construct confidence intervals using the method studied earlier in this chapter, which is for use with simple random samples only.

## EXERCISE 3.7

**1** **a** Explain what is meant by stratified random sampling.
**b** Describe the potential advantage that stratified random sampling has over simple random sampling and the population conditions for which this advantage exists.

**2** A sample survey is planned at a certain school to find out pupils' favourite TV programmes. Why might it be advantageous to employ stratified random sampling? School records show that the number of pupils enrolled in each year group is as follows:

| Year | S1 | S2 | S3 | S4 | S5 | S6 |
|---|---|---|---|---|---|---|
| Boys | 80 | 77 | 92 | 82 | 59 | 46 |
| Girls | 85 | 83 | 73 | 82 | 61 | 50 |

Write down detailed instructions on how to obtain a 10% stratified random sample at this school. How would you draw a 5% stratified random sample at your school?

**3** In a certain school, 70 S5 pupils are studying Higher Mathematics. They are organised into three classes according to their mathematics results in previous years. Part way through the Higher Mathematics course they all sit the same exam and their marks are listed here:

| **Class** | **Marks(%)** | | | | | | | | | | | | | |
|---|---|---|---|---|---|---|---|---|---|---|---|---|---|---|
| **5H1** | 50 | 65 | 60 | 42 | 64 | 46 | 59 | 59 | 64 | 72 | 62 | 59 | 76 | 53 |
| | 66 | 66 | 69 | 67 | 87 | 68 | 74 | 73 | 58 | 92 | 81 | 49 | 86 | 73 |
| **5H2** | 47 | 66 | 62 | 51 | 39 | 42 | 67 | 68 | 52 | 39 | 69 | 61 | 48 | 55 |
| | 71 | 57 | 53 | 55 | 44 | 39 | 43 | 60 | 32 | 46 | 73 | 47 | 29 | 46 |
| **5H3** | 27 | 39 | 34 | 32 | 25 | 19 | 22 | 30 | 24 | 39 | 53 | 48 | 52 | 43 |

Estimate the mean mark for this population by taking a number of different samples:
**(i)** simple random samples of size 5, **(ii)** simple random samples of size 10,
**(iii)** stratified random samples of size 5 using proportional allocation.
Share the work among members of your class.
With the aid of a multiple boxplot, compare the estimates produced by these three sampling plans. Given that, generally, the larger the sample, the greater the cost, which sampling method would you recommend and why?

## *Cluster sampling*

Suppose, for example, we wished to obtain a sample of first-year pupils in Scottish secondary schools. It would be difficult to obtain a sampling frame by listing all the individual pupils in this population. Instead, it would be easier to list all the schools (clusters). A simple random sample of schools could then be selected and all the first-year pupils within each chosen school would form the sample.

With **cluster sampling** the population is divided into non-overlapping groups of units called clusters. Every member of the population belongs to one and only one of these clusters. A simple random sample of clusters is chosen and every individual within each of the selected clusters is studied. This is sometimes referred to as **one-stage cluster sampling**.

Cluster sampling is used when it is difficult or expensive to list all the individuals in a population, but it is easier to list identifiable groups or clusters of individuals. Like stratified sampling, cluster sampling involves subdividing the population into non-overlapping groups of units. However, there are some very important differences in the nature of strata and clusters. In contrast to strata, clusters should be very similar to one another because each cluster should reflect the distribution of characteristics in the population as a whole. The following table compares the qualities that strata and clusters should have ideally.

| | Within groups | Between groups |
|---|---|---|
| Strata | Groups contain units that are alike in some way and so the variability within each stratum is less than in the population as a whole | Strata differ markedly from each other and so the variability between groups tends to be high |
| Clusters | Each group is a collection of units which represents the variety within the population as a whole and so the variability within each cluster should be similar to that in the population | Clusters are very similar to each other and so the variability between groups tends to be low |

It is hoped that the first-year pupils in the schools chosen are representative of those in Scotland generally. Although cluster sampling increases convenience, it can happen that the selected clusters (schools in this case) have features which are not typical of the population as a whole. The greater the variability between clusters, the more likely it is that the sample will be unrepresentative. Unlike stratified random sampling, there is no gain in precision when estimating population parameters by cluster sampling.

There are many ways in which this basic cluster sampling method can be modified or extended to suit more complicated situations. For example, having randomly chosen a sample of clusters, a simple random sample of the units within each cluster could be selected. This is referred to as **two-stage cluster sampling**. Occasionally it is possible to identify subgroups within clusters which may be randomly sampled and this can lead to **multi-stage cluster sampling** designs. In our example, a two-stage cluster sample can be obtained if a simple random sample of the first-year pupils is drawn from within each of the chosen schools. The sampling frame for this further stage is readily available from school records.

Cluster sampling produces estimates which are unbiased. However, unlike stratified random sampling it does not necessarily improve precision. Its main advantage is administrative convenience.

*Warning*

When using cluster sampling, do not construct confidence intervals using the method studied earlier in this chapter, which is for use with simple random samples only.

**EXERCISE 3.8**

1 **a** Explain what is meant by cluster sampling.
**b** Why might cluster sampling be used instead of simple random sampling in certain situations?
**c** State a potential disadvantage of cluster sampling.

2 Define and contrast strata and clusters of a population.

3 The table of data on family size (Appendix 3, page 142) shows the number of children in families of pupils who attend a particular school. There are 700 pupils at the school and they are organised into 24 register classes. Each column in the table gives the data for one of these register classes. The register classes have been numbered 1 to 24 as shown in the top row of the table. The second row of the table records how many pupils there are in each register class. We wish to estimate the proportion of the pupils at this school who report that they are an 'only child'.
**a** By considering register classes as clusters, randomly choose one cluster and include all the individuals in the selected register class in the sample.
**b** Estimate the proportion of one-child families from this cluster sample.
**c** Select a simple random sample of 30 individuals and calculate the sample proportion of one-child families. To do this, imagine that the data in the table of family size have been numbered from 1 to 700 (e.g. class 1 would contain numbers 1 to 33, class 2 numbers 34 to 66, and so on) and use 30 distinct random numbers in the range 1 to 700 to select which data to include in the sample.
**d** Estimate the proportion of one-child families from this simple random sample.
**e** Which method of sampling did you find more convenient? The true proportion of one-child families is $\frac{78}{700} \approx 0.111$. Comment on your sampling results.
**f** Estimate the proportion of pupils in your school who are an 'only child'.

## *Further sampling methods*

The decision to extend the M77 (Glasgow to Ayr motorway) so that it runs through the Pollock Estate (green land) in the south of Glasgow generated a lot of controversy in the late 1990s. How might we have estimated the proportion of Glasgow residents who were opposed to the construction of this road?

### *Convenience sampling*

Units that are readily available form a convenience sample. There is no random mechanism employed in the selection process. Although a convenience sample can be obtained relatively cheaply, it is not usually representative. For example, suppose that a local newspaper prints the following questionnaire in one of its issues:

| Do you favour or oppose the building of the M77 extension through Pollock Estate? | ☐ | **Favour** |
|---|---|---|
| | ☐ | **Oppose** |

The newspaper invites its readers to fill in and return this questionnaire. It is very unlikely that the results will be representative of the opinions of Glasgow residents as a whole because only those people who (i) read this newspaper and (ii) have strong views one way or the other will respond.

### *Quota sampling*

This is the method of choice for most opinion polling and market research companies. It is like stratified random sampling in making use of stratification, but differs by not using random selection. It is reasonable to believe that most young people might hold similar views about the route of the M77, but that these might be very different from the views held by older people, so age might be a reasonable stratification variable. Interviewers would be sent out to interview a given number (quota) of people in each stratum, for example, ten men aged 18–29, ten women aged 18–29, …, five men aged 70 and over, ten women aged 70 and over. As in stratified random sampling, the quotas are determined to ensure that the same sampling fraction is recruited in each stratum of the population.

The crucial difference between quota sampling and stratified random sampling is that the samples are not drawn randomly. The interviewers are placed in busy places like shopping centres, and are allowed to choose which particular passers-by to interview (as long as, overall, they meet their quotas). One possible source of bias in quota sampling is that interviewers themselves choose whom to interview within each stratum and, too often, these are people they expect to be pleasant to interview (people like themselves). Another problem is that sometimes certain strata can be under-represented because of difficulties in recruiting interviewees from those strata.

### *Systematic sampling*

In the present context (M77 extension), a sampling frame would first be drawn up. This might consist of a series of Electoral Registers for all the Parliamentary constituencies in Glasgow. All voters in the frame would then be numbered consecutively, from 1 to $N$. If a 1% sample was to be drawn, then one number between 0 and 99 would be chosen at random – say 18. The sample would consist of all the voters whose number in the sampling frame ended with the chosen digits (e.g. 000018, 000118, 000218, …). All the members of the sample would be contacted and asked to take part in the opinion poll.

This method of sampling is much more expensive than the previous two, since it requires every unit in the population to be listed and the selected voters must be contacted. Its advantage is that it usually gives a representative sample. There is, however, a possible danger that if there were patterns, trends or groupings in the list then this method might be biased.

Systematic sampling is widely used by auditors. For example, suppose a company's accounts are being inspected. Rather than check every invoice, the auditors will scrutinise a systematic sample, for example all invoices with numbers ending in certain digits.

### *Non-sampling errors*

Although interview techniques and questionnaire design are beyond the scope of this course, it is important to realise that they are topics which are of considerable interest to statisticians. This is because of what are referred to as **non-sampling errors** which can bias the results obtained from an otherwise representative sample. The main non-sampling errors are:

- **non-response**, when persons selected can not be contacted or refuse to take part, or when questions on a questionnaire are left unanswered;
- **inaccurate response**, when respondents do not tell the truth, or when poorly trained interviewers influence the respondents, or when badly worded questionnaires produce misleading responses.

*Warning*

The method for constructing confidence intervals studied earlier in this chapter is for use with simple random samples only. It should not be used with convenience samples. Sometimes it is used to construct confidence intervals with quota samples and systematic samples. However, this use requires further assumptions to be made (which may or may not be true), the justification of which lies beyond the scope of this course.

## EXERCISE 3.9

**1** A sample survey of pupils at a certain school was planned to find out their opinions on proposed improvements to the school grounds. The following sampling methods were considered.

**a** Approach pupils as they arrive at school in the morning.
**b** Choose boys and girls at random from each year's listing and in proportion to the number of boys and girls in each year.
**c** Select three register classes at random and include each pupil in these classes.
**d** Choose pupils during the morning interval, ensuring representative proportions with regard to year and sex.
**e** Begin with a randomly chosen name from the first 20 names on the school's roll and choose every 20th pupil thereafter.
**f** Number all pupils from 1 to $N$ and use random numbers in this range to select $n$ pupils.
**g** Advertise a lunchtime meeting and ask those interested to attend.

Name each of the above sampling methods and list its advantages and disadvantages.

2 Describe quota sampling and stratified random sampling. Explain the relative advantages and disadvantages of these two sampling methods, in the context of opinion polling.

3 Suppose a Local Authority has 100 000 voters on its Electoral Register.
   **a** Explain how you would draw a 2% systematic sample of voters.
   **b** Each voter in the sample is sent a questionnaire on council services and 28% of them return the completed questionnaire. Of these, 291 said they were satisfied with their refuse collection service. Construct a 95% confidence interval for the true proportion of voters who are satisfied with this service.
   **c** Why might this confidence interval give a false indication of the proportion of satisfied voters?

# STATISTICS IN ACTION – SAMPLING

## *1. Sampling distributions*

**1** To estimate the mean $\mu$ for population A:

**a** choose a sample of four components from population A by simple random sampling,

**b** calculate the sample mean of the components' diameters (answers to one decimal place),

**c** use a stem and leaf diagram to display the sample means for the whole class (about 20 results).

**2** Repeat the above procedure for population B.

**3** **a** Draw stem and leaf diagrams for populations A and B.

**b** Which population has the greater natural variation?

**c** Compare the stem and leaf diagrams of the estimates in questions 1 and 2.

**d** Did the samples from one of the populations give less variable estimates of the population's mean? Explain.

**4** **a** Estimate the mean of population B using random samples of size $n = 9$.

**b** Use a stem and leaf diagram to illustrate the results for the whole class.

**c** What effect does increased sample size have on the variability of estimates?

| Component number | Population A Diameter (mm) | Population B Diameter (mm) |
|---|---|---|
| 1 | 8.1 | 7.3 |
| 2 | 7.7 | 4.8 |
| 3 | 6.8 | 10.3 |
| 4 | 6.9 | 8.1 |
| 5 | 7.4 | 5.2 |
| 6 | 6.4 | 3.5 |
| 7 | 7.8 | 11.2 |
| 8 | 5.9 | 7.6 |
| 9 | 7.7 | 7.7 |
| 10 | 7.0 | 9.4 |
| 11 | 6.7 | 6.0 |
| 12 | 6.1 | 9.0 |
| 13 | 9.2 | 6.1 |
| 14 | 6.6 | 6.4 |
| 15 | 6.8 | 9.0 |
| 16 | 8.7 | 7.8 |
| 17 | 6.3 | 10.1 |
| 18 | 9.5 | 9.5 |
| 19 | 6.5 | 5.2 |
| 20 | 6.7 | 9.5 |

## *2. Estimating average wheat yield*

In an agricultural experiment, a large field was sown uniformly with a particular variety of wheat. The seed was sown in rows that were spaced 12 inches apart. Each row was divided into sections 15 feet long and these sections are referred to as plots of wheat. The yield of each plot was recorded as the number of grains of wheat harvested.

In Appendix 4 (page 143) the yields of 1000 plots close to the centre of the field have been tabulated. These plots were in an area consisting of 100 rows with 10 plots in each row. For convenience the rows have been numbered 00 to 99 and the plots have been numbered 0 to 9. These 1000 plots are the population for the following exercise where the population mean wheat yield per plot will be estimated from the yields of a random sample of plots.

Choose a simple random sample of 30 different plots. One way to do this is to use your calculator to generate 30 different random three-digit numbers in the range 000 to 999.

The first two digits are used to select the row and the last digit to choose the plot in that row. For example, the random number 261 would indicate row 26 plot 1, where from Appendix 4 we see that the yield was 660 grains.

Calculate the sample mean

$$\bar{x} = \frac{\Sigma x}{30}$$

and the sample standard deviation

$$s = \sqrt{\frac{\Sigma x^2 - (\Sigma x)^2/30}{29}}$$

The 95% confidence interval for the mean yield per plot of the population of 1000 plots is given by

$$\left(\bar{x} - 1.96\frac{\sigma}{\sqrt{n}}, \bar{x} + 1.96\frac{\sigma}{\sqrt{n}}\right)$$

Now the true population standard deviation $\sigma$ is unknown. A formal procedure for constructing confidence intervals when $\sigma$ is unknown will be introduced in the Advanced Higher Unit, Statistics 2.

However, an approximate 95% confidence interval can be obtained from your sample of 30 plots using the formula:

$$\left(\bar{x} - \frac{2s}{\sqrt{30}}, \bar{x} + \frac{2s}{\sqrt{30}}\right)$$

Use this quick formula to construct an approximate 95% confidence interval from your sample data.

In a real investigation of this type, the only information we would have is the yield of the 30 plots which were randomly selected in the field. However, our investigation is an artificial exercise because the complete information on the population is available in Appendix 4 and so the true population mean yield is known to be 578 grains per plot.

Does your confidence interval include this true value? Compare your answer with the answers of other members in your class (about 20 results). How many of the confidence intervals included the true value?

Another way to sample the plots is to use stratified random sampling where we consider the columns of plots as strata numbered 0 to 9. Because the natural fertility of the soil may vary across the field, stratified random sampling may improve the precision of the estimate of wheat yield by ensuring that the randomly selected plots are spread across the whole area being studied.

Use your calculator to generate three different two-digit random numbers and use these to select three rows in the column for plot 0. Repeat this process to select three rows at random from each of the columns for plot 1 through to plot 9.

Calculate the sample mean. Compare your answer with the answers of other members in your class. Use a multiple boxplot to compare the point estimates produced by simple random sampling with those produced by stratified random sampling.

### 3. Computer simulation to demonstrate the Central Limit Theorem

The following computer program has been written in QBASIC which will run on any PC. If your computer uses a different dialect of BASIC you may need to modify the program slightly. The program outputs data onto a floppy disk as an ASCII file called 'result.dat'. This file can be read by a spreadsheet or statistics package for analysis.

```
100  RANDOMIZE TIMER
110  CLS
120  PRINT "CLT Demonstration": PRINT
130  PRINT "Put a floppy disk in Drive A": PRINT
140  PRINT "Input sample size";
150  INPUT n%
160  PRINT: PRINT "How many samples";
170  INPUT samples%
180  name$ = "a:\result.dat"
190  PRINT: PRINT "Saving results to "; name$
200  OPEN name$ FOR OUTPUT AS #1
210  FOR p% = 1 TO samples%
220  total% = 0
230  FOR i% = 1 TO n%
240  total% = total% + INT(RND * 6) + 1
250  NEXT i%
260  mean = total% / n%
270  PRINT #1, mean
280  NEXT p%
900  CLOSE
910  END
```

When you run this program, you will have to

- put a floppy disk into Drive A,
- type in the sample size (press 'Return' or 'Enter' when you've done this),
- type in how many samples you wish (press 'Return' or 'Enter').

Try the following investigation.

Task 1 Run the program and use a sample size of one and ask for 100 samples.
The program simulates rolling a die 100 times.
The file 'result.dat' will contain 100 scores of the die ranging from 1 to 6.
Display these scores in a bar diagram.

Task 2 Run the program again but this time request a sample size of five and ask for 100 samples.
The program simulates rolling the die five times and calculates the mean of these five scores.
The file 'result.dat' will contain 100 sample means of five rolls of the die.
Display these 100 sample means in a suitable diagram.

Task 3 Repeat Task 2 for sample sizes **(i)** 10 **(ii)** 20 **(iii)** 50.

On one occasion this investigation produced the following results.

The bar diagram for Task 1 shows a discrete uniform distribution of the scores on the die in the simulation.

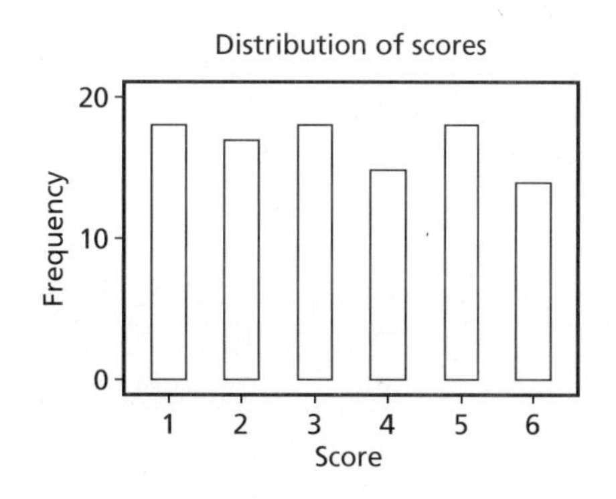


The diagrams below show that as the sample size increases the shape of a Normal curve appears and the sampling variation decreases (in line with the Central Limit Theorem).

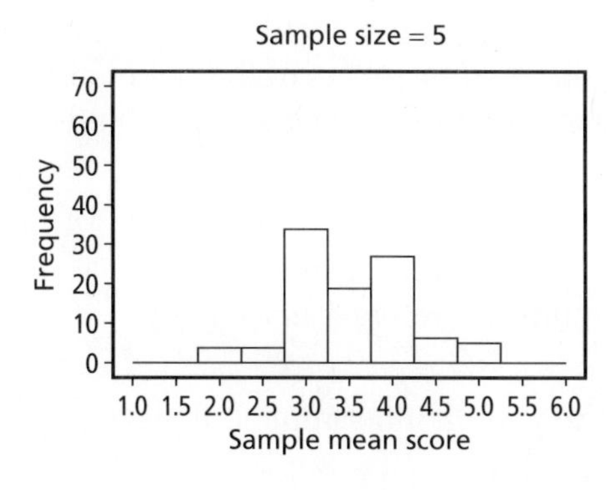


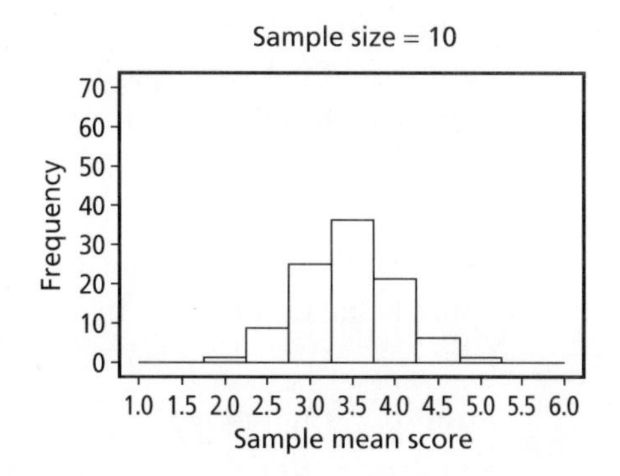


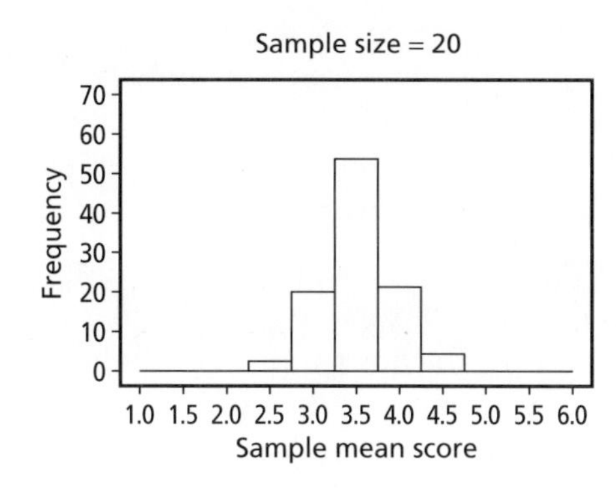


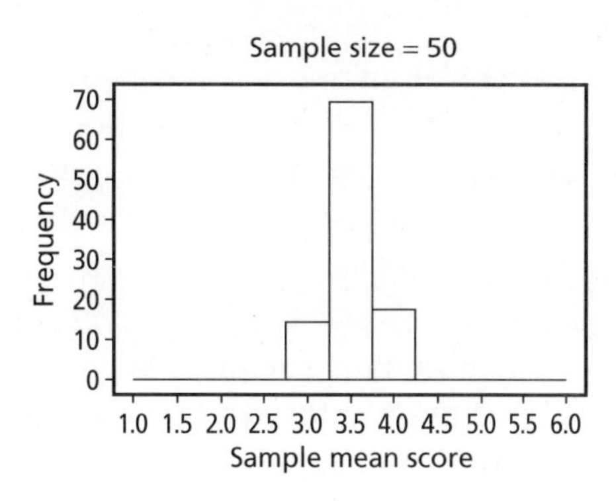

# CHAPTER 3 SUMMARY

**1** A **population** is the whole collection of individuals (people, animals, plants, or things) that we aim to study and about which we intend to draw conclusions.

**2** A **sample** is the subset of the population from which we intend to collect data. We try to ensure that the sample gives an honest representation of the population. A method of sampling which does not result in an honest representation of the population is said to be biased. This can result in wrong conclusions being drawn about the population.

**3** **Sampling units** are non-overlapping collections of individuals from the population. In simple cases, the units are individual members of the population. However, sometimes it is convenient to define larger units such as households, post-code districts, schools, etc.

**4** A **sampling frame** is an exhaustive list of all the units in the population. Care must be taken to avoid incompleteness, duplication and inaccuracy.

**5** The **sampling fraction** is the proportion of the population sampled. If a sample of size $n$ is drawn from a population of size $N$ then the sampling fraction is $\frac{n}{N}$.

**6** **Statistical inference** is the study of how to use information from a sample to draw valid conclusions about the population from which the sample was drawn. Study of the sample may produce a **sample statistic** such as the sample mean $\bar{x}$ which is used to infer the value of the corresponding **population parameter**, namely the population mean $\mu$.

**7** **Simple random sampling** is done by selecting units for the sample by numbering all the units in the sampling frame consecutively, from 1 to $N$ say, and then choosing $n$ of them without replacement, using $n$ different random numbers in the range 1 to $N$.

The number of different possible simple random samples of size $n$ is $\binom{N}{n}$ and each of these samples is equally likely to be chosen.

**8** **The distribution of sample means**

If $X_1, X_2, \ldots, X_n$ $(n \geq 2)$ are independent random variables with the same distribution, then we say that these random variables are independent and identically distributed (iid). This represents the situation when we draw a simple random sample of size $n$ from a population, where the random process used in selecting the sample allows us to assume independence.

In particular, suppose that for each $X_i$

$$E(X_i) = \mu \quad \text{and} \quad V(X_i) = \sigma^2$$

then

$$E(\overline{X}) = \mu \quad \text{and} \quad V(\overline{X}) = \frac{\sigma^2}{n}$$

Note that

$$V(\overline{X}) < V(X_i)$$

Furthermore, if

$$X_i \sim N(\mu, \sigma^2) \text{ iid}$$

then

$$\overline{X} \sim N\left(\mu, \frac{\sigma^2}{n}\right)$$

**9** **The standard error of the mean** is a numerical indicator of the precision of an estimate of the population mean.

$$\text{Standard error} = \sqrt{V(\overline{X})} = \frac{\sigma}{\sqrt{n}}$$

**10** **The Central Limit Theorem**

If $X_1, X_2, \ldots, X_n$ is a sequence of independent and identically distributed random variables, each with expected value $\mu$ and variance $\sigma^2$, then, for sufficiently large $n$,

$$\overline{X} \sim N\left(\mu, \frac{\sigma^2}{n}\right) \text{ approximately}$$

regardless of the parent distribution of the $X_i$.

The larger $n$ is, the better the approximation. The Central Limit Theorem is sometimes stated in an alternative way which is equivalent to the above, namely: for sufficiently large $n$,

$$\sum_{i=1}^{n} X_i \sim N(n\mu, n\sigma^2) \text{ approximately}$$

**11** A **confidence interval for a population mean** is a range of plausible values for the population mean. From a simple random sample, calculate

$$\overline{x} \pm k\frac{\sigma}{\sqrt{n}}$$

For 90% CI use $k = 1.645$
For 95% CI use $k = 1.96$
For 99% CI use $k = 2.58$

where the population standard deviation $\sigma$ is known.

A 95% confidence interval is such that

$$P\left(\overline{X} - 1.96\frac{\sigma}{\sqrt{n}} < \mu < \overline{X} + 1.96\frac{\sigma}{\sqrt{n}}\right) = 0.95$$

This interval will not always contain the true value of the population parameter. However, averaged over our lifetime, about 95% of the intervals we calculate will capture the true value.

**12** A **confidence interval for a population proportion** is a range of plausible values for the population proportion. From a simple random sample, calculate an estimate $\hat{p} = \frac{X}{n}$. An approximate 95% confidence interval for $p$ is given by

$$\hat{p} \pm 1.96\sqrt{\frac{\hat{p}\hat{q}}{n}} \quad \text{where } \hat{q} = 1 - \hat{p}$$

provided $n \geq 20$, $np \geq 5$, and $nq \geq 5$.

The quantity $\sqrt{\frac{\hat{p}\hat{q}}{n}}$ is known as the estimated standard error of the sample proportion.

**13** **Stratified random sampling** is used when the population can be divided into non-overlapping groups or strata, where the units in each stratum are thought to have similar characteristics. Every member of the population belongs to one and only one of these strata. There will be less variability within these strata than in the population as a whole. Stratification can improve the precision of estimation. A sampling frame is drawn up for each stratum separately and a simple random sample is then drawn from each stratum. Usually, a common sampling fraction is used in all the strata.

**14** In **cluster sampling** the population is divided into non-overlapping groups of units called clusters. Every member of the population belongs to one and only one of these clusters. A simple random sample of clusters is chosen and every individual within each of the selected clusters is studied. This is sometimes referred to as **one-stage cluster sampling**. Unlike stratified random sampling, there is no gain in precision when estimating population parameters by cluster sampling. Its main advantage is administrative convenience.

**15** In **convenience sampling** units that are readily available form a convenience sample. There is no random mechanism employed in the selection process. Although such samples can be obtained cheaply, they are not usually representative.

**16** **Quota sampling** is like stratified random sampling because it makes use of stratification, but it differs because it does not use random selection.

**17** In **systematic sampling** a sampling frame is first drawn up and all units numbered consecutively, from 1 to $N$. If a 1% sample is to be drawn, then one number between 0 and 99 would be chosen at random – say 18. The sample would consist of all the units whose number in the sampling frame ended with the chosen digits (e.g. 000018, 000118, 000218, ...). Systematic samples are usually representative.

# CHAPTER 3 REVIEW EXERCISE

**1** Describe the advantages and disadvantages of conducting a sample survey rather than a census.

**2** $X \sim N(20, 16)$. A computer program generated a random sample of four observations from this distribution:

21.2　　22.3　　12.9　　39.6

**a** Calculate the sample mean $\bar{x}$.
**b** Calculate the standard error.
**c** Determine the distribution of $\bar{X}$.
**d** Calculate **(i)** $P(X < \bar{x})$ **(ii)** $P(\bar{X} < \bar{x})$.

**3** The discrete random variables $X_i (i = 1, 2, \dots, 25)$ are iid, $X_i \sim \text{Bin}(15, 0.6)$.
**a** Write down $E(X_i)$, $V(X_i)$ and determine the approximate distribution of $T \sim \sum_{i=1}^{25} X_i$.
**b** Calculate $P(T < 250)$ and hence write down $P(\bar{X} < 10)$ where $\bar{X} = \frac{1}{25}T$.

**4** A graph of the probability density function of a continuous random variable $X$ is shown:

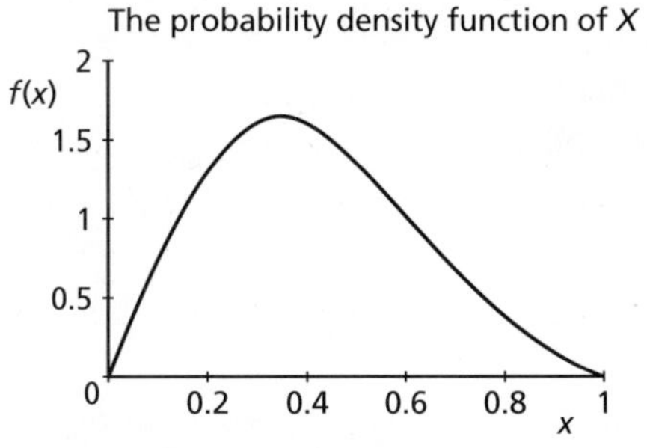

The probability density function of $X$

f(x)
2
1.5
1
0.5
0
0.2
0.4
0.6
0.8
1
x

The mean $\mu = 0.4$ and variance $\sigma^2 = 0.04$.
**a** If random samples of four observations of $X$ are taken, state the approximate sampling distribution of $\bar{X}$.
**b** Calculate $P(\bar{X} < 0.5)$.

**5** **a** Describe
**(i)** quota sampling,
**(ii)** systematic sampling,
and explain how a biased sample could arise when using each of these methods.
**b** Giving examples, explain what is meant by non-sampling errors.

**6** Compare and contrast stratified random sampling and cluster sampling.

7 The grip strength of a simple random sample of students was measured in suitable units and the following data show the results for males and females separately:

| | | | | | | |
|---|---|---|---|---|---|---|
| Males | 32 | 34 | 36 | 37 | 38 | 39 |
| | 41 | 44 | 44 | 44 | 44 | 45 |
| | 45 | 46 | 47 | 48 | 48 | 49 |
| | 50 | 51 | 53 | 53 | 55 | 55 |
| Females | 17 | 20 | 22 | 23 | 24 | 25 |
| | 25 | 26 | 26 | 28 | 28 | 28 |
| | 29 | 29 | 29 | 30 | 31 | 33 |
| | 33 | 34 | 35 | 36 | 36 | 37 |

**a** Display these data in a multiple boxplot.
**b** Calculate the sample mean for **(i)** male, **(ii)** female students.
**c** Given that the population standard deviation for males is $\sigma = 6$ and that $\sigma = 5$ for females, calculate the standard error for **(i)** males, **(ii)** females.
**d** Construct a 95% confidence interval for the population mean grip strength of **(i)** males, **(ii)** females.
**e** Compare the grip strength of male and female students.

8 A simple random sample of 100 books was selected from a school library and for each book the number of times it had been borrowed during the past year was counted. The following data were recorded:

| Number of times borrowed | Frequency (number of books) |
|---|---|
| 0 | 12 |
| 1 | 53 |
| 2 | 22 |
| 3 | 10 |
| 4 | 3 |

**a** Calculate the sample proportion of books borrowed more than once.
**b** Calculate the estimated standard error.
**c** Construct an approximate 95% CI for the true proportion of books borrowed more than once from the library. Interpret this interval carefully.

# 4 Hypothesis Testing

## *What is hypothesis testing?*

We have already seen that statistics is about drawing valid conclusions concerning a population from a sample. There is a very general requirement to do this, so statistical methods are used in most sciences and social sciences. Well-conducted research studies in these subjects begin with a question about a population. Here are some examples.

1. What is the average height of children in the west of Scotland when they start primary school?
2. Is the average environmental level of radioactivity greater in the vicinity of a nuclear installation than it is elsewhere?
3. If we put overweight people on a high-fibre diet, will they lose weight on average?
4. In 1972, 44% of adults in Scotland had none of their natural teeth. In 1988, the figure had fallen to 26% and in 1998 to 18%. Has this figure fallen further since 1998?

Another way to look at this is to say that we wish to investigate the plausibility of an **hypothesis** (or statement) about the population. Here are some possible hypotheses.

1. When children in the west of Scotland start primary school, their average height is different from that of children of the same age in the south of England (1.06 metres).
2. The average environmental level of radiocarbon in the vicinity of a nuclear installation is greater than 257 becquerel per kilogram of carbon, which is the global environmental level.
3. Very obese people who change to a high-fibre diet will lose weight on average.
4. The proportion of adults in Scotland who have none of their natural teeth is now less than 18%.

These are all statements about what researchers expect to find out. They are usually saying that something has changed, or is different, or can be made different. These are **study hypotheses**, often denoted $\mathbf{H_1}$. Hypotheses are usually written in terms of population parameters.

1. $H_1$: $\mu \neq 1.06$
   where $\mu$ (metres) is the population average height of children in the west of Scotland when they start primary school.
2. $H_1$: $\mu > 257$
   where $\mu$ (becquerel per kilogram of carbon) is the population average level of radiocarbon in the vicinity of a nuclear installation.

3 $H_1$: $\mu_D > 0$
where $\mu_D$ (kilograms) is the population average weight loss that would be achieved after four weeks on a high-fibre diet.

4 $H_1$: $p < 0.18$
where $p$ is the population proportion of adults in Scotland who have none of their natural teeth.

We use the data collected from the sample to assess the evidence in favour of the study hypothesis. It is necessary also to consider what would be true if the study hypothesis were not true. We can frame this statement about the population as an hypothesis, called the **null hypothesis ($H_0$)**. $H_0$ usually represents a situation in which things have stayed the same, or two things are equal.

1 $H_0$: $\mu = 1.06$ (the mean height of children starting primary school in the south of England)

2 $H_0$: $\mu = 257$ (the global environmental level of radiocarbon)

3 $H_0$: $\mu_D = 0$ (a high-fibre diet makes no difference, on average, to weight)

4 $H_0$: $p = 0.18$ (the proportion of adults with no natural teeth has remained the same)

On the basis of the evidence provided by the sample data, we must decide which of the null and study hypotheses is more plausible. For this reason, the study hypothesis ($H_1$) is often called the **alternative hypothesis**, which is the terminology we will adopt in this book.

In the examples discussed in this book, the null hypothesis will always state that a population parameter is *equal* to some value. This is a good way to identify which of the possible hypotheses is the null hypothesis and which is the alternative hypothesis. (The null and alternative hypotheses stated for some of the above studies do not exhaust all the possibilities. For example, people on a diet might gain weight, on average, and this possibility does not come into either of the stated hypotheses. The approach we are taking can be shown to be formally equivalent to an approach which specifically allows for such possibilities in the null hypothesis, but that is beyond the scope of this course.)

It might seem as though we should treat the two hypotheses equally, but we do not. Null hypotheses usually, by their nature, cannot be proved to be true unless we have access to data from the whole population.

For example, suppose you decided to conduct a study with the study hypothesis that 'Some polar bears are not white'. Then, the null hypothesis is 'All polar bears are white'. If you observe a small sample of polar bears, and just one is brown, then you have disproved $H_0$ and proved $H_1$. On the other hand, even if you search for many years, and observe thousands of white polar bears, you have still not proved that $H_0$ is true. There might still be a polar bear somewhere, unobserved by you, that is not white. You could not prove $H_0$ unless you observed every single polar bear in the population, and found that every one was white. So $H_0$ cannot be proved from a sample.

When assessing the evidence from a sample, then, we have just two possibilities. Either we **reject $H_0$ in favour of $H_1$** (we have proved beyond reasonable doubt that $H_1$ is true), or we **do not reject $H_0$**. We can never prove $H_0$, though often we might not have sufficient evidence to reject it.

## EXERCISE 4.1

Write down appropriate null and alternative hypotheses corresponding to each of the following research questions.

**1** In 1998, 74% of Scottish adults brushed their teeth at least twice each day. Has this proportion increased since then?

**2** A certain type of crisp is sold in packs that nominally weigh 25 g. On average, do packs weigh less than the nominal amount?

**3** A machine is set to produce metal rods of cross-sectional diameter 5 mm. On average, is the machine producing rods that are too wide?

**4** Extensive testing with a particular spatial memory task has shown that Primary 7 pupils achieve an average score of 22 out of 30. On average, do Secondary 2 pupils achieve a different score?

**5** An exercise programme has been developed to improve young women's cardiovascular fitness. If females aged 20 to 29 years old could have their fitness measured on a treadmill test before and after four weeks on this exercise programme, would their average cardiovascular fitness be found to improve?

**6** The average Intelligence Quotient (IQ) in the general population is 100. Is the average IQ of prisoners who have been convicted of theft different from this normal value?

**7** Historical records show that a certain manufacturing facility produces defective items at the rate of 4.5%. The facility is extensively upgraded. Is the proportion of defectives reduced as a result?

**8** Suppose teenagers with severe asthma could be given training to improve their skills in managing their own asthma. If the number of hospital admissions for each individual could be recorded for one year before and one year after the training, would the average number of hospital admissions fall?

## A test of the population mean

In a study designed to investigate people's ability to estimate length, a school pupil was asked to cut a 20 cm length of string from a ball of string, without the aid of any measuring instrument. The piece of string cut by the pupil was removed from sight and the pupil was asked to cut another piece of string, also of length 20 cm. This was repeated until the pupil had cut ten pieces of string, all intended to have length 20 cm. The lengths (in cm) of the ten pieces of string were then measured, and are given below:

20.1 19.4 23.1 19.9 22.1 20.8 23.8 21.9 23.0 20.5

Is there evidence to show that, when asked to cut a piece of string of length 20 cm, this pupil would cut off an average length that is not equal to 20 cm?

### Subjective impression

The boxplot shows that the lower quartile of the sample is just over 20 cm, in other words about $\frac{3}{4}$ of the pupil's attempts were longer than 20 cm. The sample median is over 21 cm. It seems plausible that this pupil is cutting off lengths of string whose mean length is somewhat greater than 20 cm.

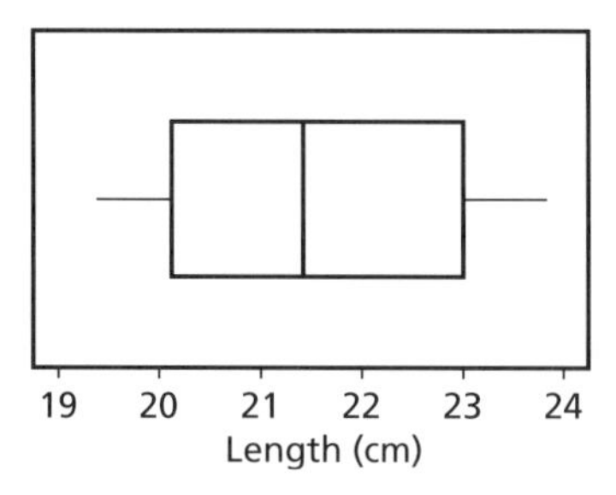


### Assumptions

We are now going to make three important assumptions.

1. All the sample values are *independent*. This seems reasonable, given the design of the experiment. For example, the pupil was given no feedback in between attempts.
2. The lengths of string cut by the pupil are *Normally distributed*. The boxplot is reasonably symmetric around the sample median, suggesting that this is a plausible assumption. (As we shall see later, the Central Limit Theorem allows us to relax the assumption of Normality for sufficiently large samples.)
3. The *population standard deviation* (or variance) of the lengths that this pupil would cut when asked to cut lengths of 20 cm is *known*. This seems unlikely to be true here, but extensive tests on school pupils of the same age suggest that a standard deviation of 1.5 cm is a reasonable value to use. (The assumption that the population standard deviation is known will be relaxed in the Advanced Higher Unit, Statistics 2.)

### Hypotheses

Let $\mu$ (cm) be the average length of string the pupil would cut under these conditions.

Then $H_0: \mu = 20 \ (= \mu_0)$

$H_1: \mu \neq 20$

These are still the hypotheses with which we started. Before looking at the data, we did not have any reason to believe that the pupil was tending to cut lengths of string longer than 20 cm, rather than shorter than 20 cm. So it would be quite wrong, on the basis of a first look at the data, to bias our investigation by substituting the alternative hypothesis that $\mu > 20$.

### *Test statistic*

It would be simplest if we were able to assess the evidence for and against $H_0$ and $H_1$ through a single function of the sample data. A function of a sample of data is known as a **statistic**; for any particular sample of data, such a function takes one and only one value, but this value varies from sample to sample. So a statistic is a random variable.

The **test statistic** we are looking for is one that has a known probability distribution when $H_0$ is true. In this case, we have assumed that the lengths of string cut by the pupil, $X_1, \ldots, X_n$, are independent $N(\mu, \sigma^2)$ random variables (where the sample size is $n = 10$ and the population standard deviation is $\sigma = 1.5$). Therefore, the sample average $\overline{X} \sim N\left(\mu, \frac{\sigma^2}{n}\right)$.

When $H_0$ is true, then

$$\overline{X} \sim N\left(\mu_0, \frac{\sigma^2}{n}\right)$$

and

$$Z = \frac{\overline{X} - \mu_0}{\sigma/\sqrt{n}} = \frac{\overline{X} - \mu_0}{\text{standard error of } \overline{X}} \sim N(0, 1)$$

The Central Limit Theorem, introduced in Chapter 3, assures us that these results also hold approximately for large enough $n$, even when the original data are not Normally distributed.

$Z$ is a suitable test statistic, since we know its distribution under $H_0$. The distribution of the test statistic under the null hypothesis, called the **null distribution**, is the Standard Normal distribution in this case. When $H_0$ is true, we expect $\overline{X}$ to be close to its hypothesised value, $\mu_0$. In other words, we expect $\overline{X} - \mu_0$ to be close to 0, and hence $Z$ to be close to 0 (its expected value when $H_0$ is true). On the other hand, when $H_1$ is true, we expect $\overline{X}$ to be further away from the hypothesised value $\mu_0$ and hence $Z$ to be further away from 0 as well.

This is the basis on which we will decide whether or not $H_0$ seems plausible given the sample data. Values of $Z$ that are close to 0 are consistent with $H_0$ rather than $H_1$. Values of $Z$ that are 'too far away' from 0, either positive or negative, are consistent with $H_1$ rather than $H_0$.

### *Observed value of the test statistic*

In this example, $\mu_0 = 20$, $\sigma = 1.5$, $n = 10$. From the data, we can easily calculate that $\Sigma x_i = 214.6$ and $\overline{x} = 21.46$. The observed value of the test statistic is

$$z = \frac{\overline{x} - \mu_0}{\sigma/\sqrt{n}} = \frac{21.46 - 20.00}{1.5/\sqrt{10}} = 3.08$$

Looking at Table 3 (Appendix 1, page 139), we can see that this is quite a large value for a Standard Normal random variable. It is quite far away from 0. If $H_0$ were true, then we would only rarely observe such a value for $Z$. It seems unlikely, given this sample of data, that $H_0$ is true. The balance of evidence from this sample suggests that we should reject $H_0$ in favour of $H_1$.

## EXERCISE 4.2

1 A common experiment in psychology involves asking a number of people to estimate the length of a given line. In one particular experiment of this kind, a line of length 135 cm was displayed on a screen. Each person in turn stood 5 metres away from the screen and guessed the length of the line. Twenty people experienced in judging distance were given this test. Their estimates of length (cm) are shown below.

| | | | | | | | | | |
|---|---|---|---|---|---|---|---|---|---|
| 119 | 124 | 125 | 125 | 128 | 130 | 130 | 130 | 130 | 130 |
| 131 | 132 | 132 | 133 | 133 | 134 | 136 | 136 | 136 | 140 |

$\Sigma x_i = 2614$

The population standard deviation of estimated length in these conditions is 5 cm.

**a** Produce a boxplot of the data. Comment on whether these experts appeared to guess the length of the line correctly on average. Check informally the assumption that these data come from a Normal distribution.

**b** Frame suitable hypotheses. Find the value of an appropriate test statistic.

2 The following data are the logarithms of fasting blood glucose levels (measured in units of mg/dl) that were determined from a sample of 25 male diabetic patients.

| | | | | | | | | | |
|---|---|---|---|---|---|---|---|---|---|
| 4.0 | 4.1 | 4.2 | 4.3 | 4.3 | 4.5 | 4.7 | 4.8 | 4.8 | 4.8 |
| 4.9 | 4.9 | 4.9 | 5.0 | 5.0 | 5.0 | 5.0 | 5.0 | 5.1 | 5.1 |
| 5.1 | 5.5 | 5.5 | 5.5 | 5.7 | | | | | |

$\Sigma x_i = 121.7$

Extensive testing in other patients suggests that the population standard deviation of (log) fasting blood glucose levels in male diabetic patients is about 0.5.

**a** Produce a boxplot of these data. Comment on the study hypothesis, that the average (log) blood glucose level in male diabetic patients is greater than 4.7 (which is the average value in the general male population). Check informally the assumption that these data have been sampled from a Normal distribution.

**b** Frame suitable hypotheses. Find the value of an appropriate test statistic.

3 **a** UK national growth standards suggest that the length of new-born baby boys is 51.09 cm on average, with a standard deviation of 2.02 cm. The lengths of new-born babies are believed to be Normally distributed. In a recent study, a random sample of 44 baby boys born at the Queen Mother's Hospital, Glasgow, were carefully measured at birth. The sample average length was 51.97 cm. Evaluate an appropriate test statistic to investigate the hypothesis that, in Glasgow, the average length of a new-born baby boy is different from that suggested by the national growth standards.

**b** According to the same growth standards, the length of new-born baby girls is 50.21 cm on average (with a standard deviation of 1.88 cm). A random sample of 41 baby girls born at the Queen Mother's Hospital had an average length of 51.11 cm. Repeat part **a** for new-born baby girls in Glasgow.

## *Assessing the evidence using a p-value*

We will continue with the 'length of string' example from the previous section, now using probability theory to give us an objective way to decide whether the observed value of the test statistic, $z = 3.08$, is far enough away from 0 to enable us to reject $H_0$ in favour of $H_1$. We have already seen that 3.08 is large in comparison with the values that we would expect to observe when $H_0$ is true, but we need an objective way of determining whether or not it is 'large enough'.

### *p-value*

One measure of how close or far away the observed value, $z$, is from zero is a probability, usually called the ***p*-value**. The $p$-value of this hypothesis test is the *conditional* probability of recording a value of $Z$ that is at least as far away from 0 as the observed $z$ *given* that $H_0$ is true. The idea behind this is that, if we decide to reject $H_0$ on the basis of the observed value, $z$, then we would also logically have to reject $H_0$ on the basis of any more extreme value.

In this example, $Z$ is at least as far away from 0 as the observed value if either $Z \geq 3.08$ or $Z \leq -3.08$.

So
$$
\begin{aligned}
p &= P(Z \geq 3.08 | H_0) + P(Z \leq -3.08 | H_0) \\
&= 2P(Z \leq -3.08 | H_0) \\
&= 2\{1 - \Phi(3.08)\} \\
&= 0.002
\end{aligned}
$$

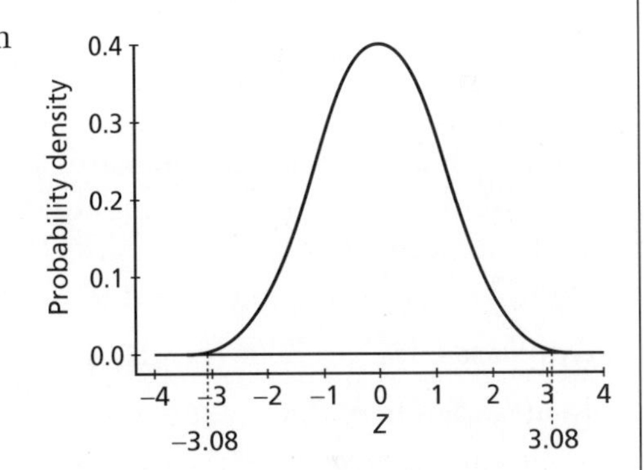


### *Conclusion*

This value of $p$ is very low. Under $H_0$, we are extremely unlikely to observe a value of the test statistic that is at least as extreme as the observed value of 3.08. So the data are not consistent with the null hypothesis. We **reject** the hypothesis that this pupil is cutting string at an average length of 20 cm, and conclude that the pupil is cutting string at an average length that is **significantly** different from 20 cm. Given the sample data, we conclude that this pupil is cutting lengths of string that are on average significantly greater than 20 cm.

Note the use of the word 'significantly' when drawing conclusions from the hypothesis test. If we had not rejected $H_0$, then we would have concluded that the pupil's average length of cut was **not significantly** different from 20 cm. This important word is used to emphasise that the conclusion is a statistical one, based on a sample of data. There is still the possibility of error – we could reject $H_0$ when $H_0$ is really true, or fail to reject $H_0$ when $H_1$ is really true. The fact that the evidence available from the sample has reached, or failed to reach, statistical significance does not guarantee that the population is really as indicated in our conclusion.

The type of alternative hypothesis used above

$$H_1: \mu \neq \mu_0$$

is known as a **two-tailed hypothesis**. The approach we used for the previous hypothesis test has to change a little when we wish instead to test a **one-tailed** alternative hypothesis, *either*

$$H_0: \mu = \mu_0 \quad \text{vs} \quad H_1: \mu < \mu_0$$

*or*, in other circumstances,

$$H_0: \mu = \mu_0 \quad \text{vs} \quad H_1: \mu > \mu_0$$

In the first case, we would reject $H_0$ only for values of $Z$ that are sufficiently small, so

$$p = P(Z \leq z | H_0) = \Phi(z)$$

In the second case, we would reject $H_0$ only for values of $Z$ that are sufficiently large, so

$$p = P(Z \geq z | H_0) = 1 - \Phi(z)$$

*Example* The daily yield of a chemical manufactured at a certain chemical plant was recorded for 50 consecutive days. The sample mean yield on these 50 days was found to be 871 tonnes. Test the alternative hypothesis that the average daily yield of chemical is less than its historical value of 880 tonnes. You may assume that the daily yield from this plant has standard deviation 21 tonnes.

*Solution*

*Assumptions.* Assume that the daily yields of chemical have mean $\mu$ tonnes and standard deviation 21 tonnes, and that the yields on different days are independent. We do not know that the yields are Normally distributed, but we rely on the Central Limit Theorem to assume that $Z \sim N(0, 1)$ approximately.

*Hypotheses.* $H_0: \mu = 880 \quad \text{vs} \quad H_1: \mu < 880$

*Observed value of Z.* $\mu_0 = 880$, $\sigma = 21$, $n = 50$, $\bar{x} = 871$, and so

$$z = \frac{\bar{x} - \mu_0}{\sigma/\sqrt{n}} = \frac{871 - 880}{21/\sqrt{50}} = -3.03$$

*p-value.* Only small values of the test statistic are consistent with $H_1$, so

$$p = P(Z \leq -3.03 | H_0) = \Phi(-3.03) = 1 - \Phi(3.03) = 1 - 0.9988 = 0.0012$$

*Conclusion.* This value of $p$ is very small, so the data are not consistent with $H_0$. We reject $H_0$ in favour of $H_1$, and conclude that the chemical plant has been producing significantly less than 880 tonnes of chemical, on average, each day.

## EXERCISE 4.3

**1** Referring back to question 1 of Exercise 4.2, on estimates of line length, now complete the hypothesis test.

**2** Referring back to question 2 of Exercise 4.2, on fasting blood glucose levels of male diabetic patients, now complete the hypothesis test.

**3** Referring back to question 3 of Exercise 4.2, on the lengths of new-born babies in Glasgow, now complete the hypothesis tests for **a** boys, **b** girls.

**4** Twenty sixth-year pupils, 10 girls and 10 boys, were independently invited to carry out a simple sorting task. After a period of training, their times to complete this task (in seconds) were as shown below.

| | | | | | | | | | | |
|---|---|---|---|---|---|---|---|---|---|---|
| Girls | 58 | 52 | 102 | 106 | 33 | 58 | 54 | 127 | 44 | 49 |
| Boys | 53 | 65 | 32 | 63 | 60 | 123 | 40 | 49 | 101 | 45 |

**a** Produce a boxplot of these data. Comment on the study hypothesis that for both boys and girls the average time required to complete this task is greater than 75 seconds. Check informally the assumption that the time required to complete the task is Normally distributed in the populations of boys and girls.

**b** Carry out an appropriate hypothesis test for boys and girls separately. In both cases, assume that the population standard deviation of the time required to complete the task is 30 seconds.



## *Assessing the evidence using a critical region*

An alternative approach to checking for statistical significance requires us to decide in advance how small the *p*-value would have to be for us to reject $H_0$ in favour of $H_1$. This pre-determined value is known as the **significance level** of the test, often denoted by the Greek symbol $\alpha$ (pronounced 'alpha'), where $\alpha$ must lie in the range 0 to 1. We reject $H_0$ for values of the test statistic $Z$ that would give a *p*-value less than $\alpha$. This means that $\alpha$ has to be a small value, so $\alpha = 0.05$ (or 5%) and $\alpha = 0.01$ (or 1%) are commonly used significance levels.

We begin by working out the value of $Z$ which gives a *p*-value exactly equal to $\alpha$. This is known as the **critical value** of the test. Any value of $Z$ that is more extreme than the critical value is associated with a *p*-value that is less than $\alpha$. So we reject $H_0$ only if the observed value of the test statistic, $z$, is more extreme than the critical value. This means that we can conduct the test without actually working out the *p*-value, since we just need to compare $z$ with the pre-determined critical value.

Suppose that we decide in advance on a significance level, $\alpha$. When testing a two-tailed alternative hypothesis, we will reject $H_0$ for values of $Z$ that are large and positive but also for values of $Z$ that are large and negative.

The critical value has to be positioned as shown on the diagram, where the shaded region indicates the associated significance level, $\alpha$.

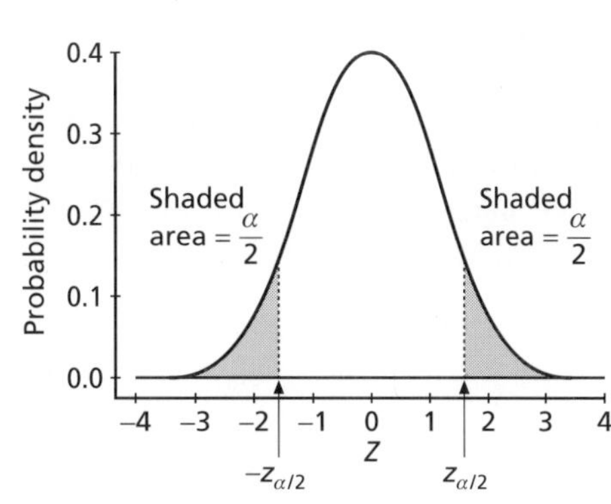


This means that the critical value is defined by the equation:

$$P(Z > \text{critical value} \mid H_0) = \frac{\alpha}{2}$$

i.e. $$P(Z > \text{critical value} \mid Z \sim N(0, 1)) = \frac{\alpha}{2}$$

The (critical) value of $Z$ that makes this equation hold is often denoted $z_{\alpha/2}$.

Values of $z_p$, for various values of $p$, are given in Table 4 (Appendix 1, page 140).

It is clear from the above diagram that, when $H_0$ is true, there is probability $\alpha$ of $Z$ being more extreme than the critical value for the test. So, when $H_0$ is true, there is probability $\alpha$ that $H_0$ is rejected. The significance level, $\alpha$, is the probability of *wrongly* rejecting $H_0$.

For a two-tailed test with significance level 0.01, the required critical value is $z_{0.005}$. Table 4 shows that $z_{0.005} = 2.58$. So, at a significance level of 0.01, we would reject $H_0$ in this two-tailed test if

$$z > z_{0.005} = 2.58 \quad \text{or} \quad z < -z_{0.005} = -2.58$$

The set of values

$$CR = \{z: z < -2.58 \ \textit{or} \ z > 2.58\}$$

is known as **the critical region** for the test. We reject $H_0$ in favour of $H_1$ if and only if the observed value of the test statistic, $z$, lies within the critical region.

In the previous example on cutting string, suppose that we adopt a fixed significance level of 0.01. Then the critical region is $CR = \{z: z > 2.58 \ \textit{or} \ z < -2.58\}$. We have already calculated that $z = 3.08$ in this example, i.e. $z$ lies in CR. So, we reject $H_0$ at a significance level of 0.01 and conclude (as before) that the mean length of string cut by this pupil in these conditions is significantly different from 20 cm. Of course, since we had already worked out that $p = 0.002 < 0.01$, then we already knew that we had to come to this conclusion!

Again, this procedure has to be amended somewhat when dealing with one-tailed alternative hypotheses. In these cases, the forms of the appropriate critical regions are:

$$CR = \{z: z < -z_\alpha\} \quad \text{when } H_1: \mu < \mu_0$$
$$CR = \{z: z > z_\alpha\} \quad \text{when } H_1: \mu > \mu_0$$

*Example* 100 rivets are sampled off a production line and measured. Their average cross-sectional diameter is found to be 6.1 cm. The nominal cross-sectional diameter of the rivets is 6 cm, and historically rivets have been produced with a standard deviation in cross-sectional diameter of 0.3 cm. Investigate the alternative hypothesis that rivets are now being produced with an average cross-sectional diameter that is greater than the nominal value.

*Solution*

*Assumptions.* Assume that the cross-sectional diameters of these rivets are Normally distributed with mean $\mu$ cm and standard deviation $\sigma = 0.3$ cm, and that the cross-sectional diameters of different rivets are independent.

*Hypotheses.* $H_0: \mu = 6$ vs $H_1: \mu > 6$

*Significance level.* Choose $\alpha = 0.05$.

*Critical region.* Large values of the test statistic are consistent with $H_1$, so

$$CR = \{z: z > z_{0.05}\} = \{z: z > 1.64\}$$

*Observed value of Z.* $\mu_0 = 6$, $\sigma = 0.3$, $n = 100$, $\bar{x} = 6.1$, and so

$$z = \frac{\bar{x} - \mu_0}{\sigma/\sqrt{n}} = \frac{6.1 - 6.0}{0.3/\sqrt{100}} = 3.33$$

*Conclusion.* The observed value of $z$ lies in the critical region, so we reject $H_0$ in favour of $H_1$. We conclude that the rivets now being produced have an average cross-sectional diameter that is significantly greater than the nominal value of 6 cm.

## EXERCISE 4.4

**1 a** Referring back to question 1 of Exercise 4.2, on estimates of line length, complete the hypothesis test with a fixed significance level of 0.05.

**b** Compare your results from part **a** above with your previous results from question 1 of Exercise 4.3.

**2 a** Referring back to question 2 of Exercise 4.2, on blood glucose levels of male diabetic patients, complete the hypothesis test with a fixed significance level of 0.01.

**b** Compare your results from part **a** above with your previous results from question 2 of Exercise 4.3.

**3 a** Referring back to question 4 of Exercise 4.3, on times required to carry out a sorting task, complete the hypothesis tests for boys and girls with fixed significance levels of 0.05.

**b** Compare your results from part **a** above with your previous results from question 4 of Exercise 4.3.

**4** A supermarket's brand of potato crisps is sold in packets with a nominal weight of 25 grams. The contents of a sample of 50 packets were weighed (to the nearest 0.1 g). The results are shown in the stem and leaf diagram below.

| Stem | Leaves |
|---|---|
| 24 | 1 4 5 7 8 |
| 25 | 0 2 2 2 3 3 5 5 5 6 7 7 8 8 8 8 8 |
| 26 | 0 0 0 0 1 1 2 2 3 3 3 3 4 4 5 6 7 9 |
| 27 | 0 0 1 1 3 5 6 7 8 |
| 28 | 5 |

24 | 1 represents 24.1 g

The weight of packets of crisps like these has a standard deviation of 1 g. Using a fixed significance level of 0.001, investigate the hypothesis that the average weight of packets of crisps like these is greater than 25 g.

## Paired data

Charles Darwin designed the following experiment to investigate the study hypothesis that cross-fertilised plants are more vigorous than self-fertilised plants. Fifteen pairs of seedlings of the same age, one of them generated by cross-fertilisation and the other by self-fertilisation, were grown together so that the members of each pair were grown under identical conditions. The heights of the plants (in inches) after a fixed period of time are shown below.

| Pair, $i$ | Height of cross-fertilised plant, $x_i$ | Height of self-fertilised plant, $y_i$ | Difference in heights, $d_i = x_i - y_i$ |
|---|---|---|---|
| 1 | 23.5 | 17.4 | 6.1 |
| 2 | 12.0 | 20.4 | –8.4 |
| 3 | 21.0 | 20.0 | 1.0 |
| 4 | 22.0 | 20.0 | 2.0 |
| 5 | 19.1 | 18.4 | 0.7 |
| 6 | 21.5 | 18.6 | 2.9 |
| 7 | 22.1 | 18.6 | 3.5 |
| 8 | 20.4 | 15.3 | 5.1 |
| 9 | 18.3 | 16.5 | 1.8 |
| 10 | 21.6 | 18.0 | 3.6 |
| 11 | 23.3 | 16.3 | 7.0 |
| 12 | 21.0 | 18.0 | 3.0 |
| 13 | 22.1 | 12.8 | 9.3 |
| 14 | 23.0 | 15.5 | 7.5 |
| 15 | 12.0 | 18.0 | –6.0 |

Did Darwin's experiment succeed in demonstrating his study hypothesis? Unlike previous examples, in this case we have *two* populations of interest: the population of cross-fertilised plants and the population of self-fertilised plants. Let $\mu_X$ and $\mu_Y$, respectively, denote the population mean heights of the cross-fertilised and self-fertilised plants. Then Darwin was interested in the following hypotheses:

$$H_0: \mu_X = \mu_Y \quad \text{vs} \quad H_1: \mu_X > \mu_Y$$

(Note the one-tailed alternative hypothesis.)

It is possible to test hypotheses like these directly, using independent samples from the two populations, but that type of test is beyond the scope of the present course. (See the Advanced Higher Unit, Statistics 2, for further details.) The design of this experiment makes it possible, however, to evaluate the evidence for and against these hypotheses using the $z$ test we have already developed for one-sample problems.

To do this, we need to reframe the hypotheses in terms of the differences in means,

$$\mu_D = \mu_X - \mu_Y$$

The hypotheses become

$$H_0: \mu_D = 0 \quad \text{vs} \quad H_1: \mu_D > 0$$

Now we need to decide how to find numerical data that would help us find out directly about the value of $\mu_D$. The **paired differences**, $d_i = x_i - y_i$, clearly fill this role. Since $x_i$ and $y_i$ are measurements of the vigour of a cross-fertilised and a self-fertilised plant grown under identical conditions then, the larger $x_i$ is in comparison with $y_i$, the more vigorous was that cross-fertilised plant relative to the self-fertilised plant grown under the same conditions.

$d_1, d_2, \ldots, d_n$ are measurements from a population that has mean value $\mu_D$. Again, if we can assume that these paired differences are independent observations from a Normal distribution, with known standard deviation $\sigma_D$, then we can use the $z$ test developed previously to test these hypotheses.

*Subjective impression.* This boxplot of the paired differences suggests that the data are reasonably symmetric around the sample median, though there is one very low outlier (a pair in which the self-fertilised plant is a lot taller than the cross-fertilised plant). More than $\frac{3}{4}$ of the paired differences are positive, indicating that cross-fertilised plants are generally more vigorous than self-fertilised plants grown under identical conditions.

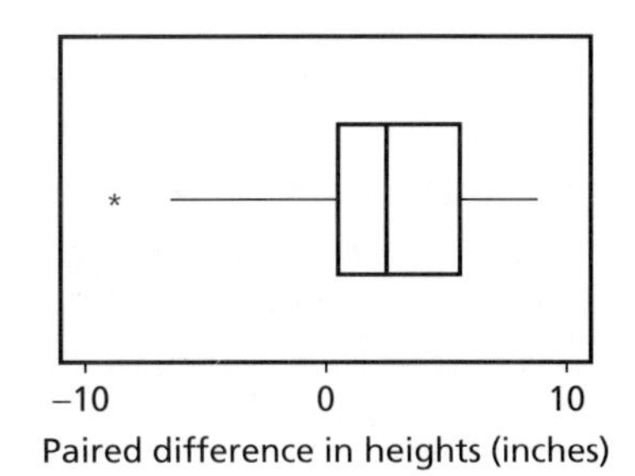


*Assumptions.* Assume that the paired differences are Normally distributed with mean $\mu_D$ inches and standard deviation 5 inches (a figure based on previous experience). Assume that the data from different pairs are independent.

*Hypotheses.* $H_0: \mu_D = 0$ vs $H_1: \mu_D > 0$

*Observed value of Z.* $\mu_0 = 0$, $\sigma_D = 5$, $n = 15$ (the number of *pairs*), $\bar{d} = 2.61$ and so

$$z = \frac{\bar{d} - \mu_0}{\sigma_D/\sqrt{n}} = \frac{2.61 - 0}{5/\sqrt{15}} = 2.02$$

*p-value.* Only large values of the test statistic are consistent with $H_1$, so

$$p = P(Z \geq 2.02 \mid H_0) = 1 - \Phi(2.02) = 1 - 0.9783 = 0.0217$$

*Conclusion.* This value of $p$ is small, so the data are not consistent with $H_0$. We reject $H_0$ in favour of $H_1$, and conclude that cross-fertilised plants are, on average, more vigorous than self-fertilised plants grown under identical conditions.

## EXERCISE 4.5B

1 An experiment involving a simple sorting task was introduced in question 4 of Exercise 4.3. The data given there refer to the pupils' third attempt at the task, following a period of training. The table below shows the time required (in seconds) for each pupil's first and third attempts.

| Girls | | | | | | | | | | |
|---|---|---|---|---|---|---|---|---|---|---|
| First attempt | 87 | 64 | 110 | 148 | 65 | 106 | 107 | 159 | 85 | 98 |
| Third attempt | 58 | 52 | 102 | 106 | 33 | 58 | 54 | 127 | 44 | 49 |
| **Boys** | | | | | | | | | | |
| First attempt | 107 | 63 | 67 | 100 | 119 | 129 | 103 | 75 | 101 | 71 |
| Third attempt | 53 | 65 | 32 | 63 | 60 | 123 | 40 | 49 | 101 | 45 |

a Calculate the paired difference between the times required for each girl's first and third attempts at the task, and display these paired differences on a boxplot. Comment on the assumption that these values are a sample from a Normal distribution. Investigate the hypothesis that, for girls, the average reduction in the time required to complete this task is more than 15 seconds. (You may assume that the reduction in time required has a population standard deviation of 20 seconds.)

b Repeat part **a** for boys. (You may again assume that the population standard deviation is 20 seconds.)

2 Recently, 18 children suffering from cystic fibrosis (CF) caught an infection that can cause a sudden deterioration in lung function. Each child's forced expiratory volume in 1 second (FEV1) was recorded in routine tests shortly before the outbreak of this infection. Higher values of FEV1 indicate better lung function. FEV1 was measured again some months after the outbreak. The recorded change in each child's FEV1 level, where change = later level – initial level, are shown below (expressed as a percentage of the normal value for children of the same sex and height).

| | | | | | | | | |
|---|---|---|---|---|---|---|---|---|
| −63 | −34 | −25 | −23 | −20 | −18 | −17 | −13 | −12 |
| −12 | −5 | −4 | −2 | +4 | +17 | +21 | +35 | +43 |

a Display these data on a boxplot. Comment on the study hypothesis that, on average, FEV1 decreases as a result of contracting this infection. Comment on the assumption that these data are a sample from a Normal distribution.

b Carry out an appropriate hypothesis test to investigate the study hypothesis. (You may assume that the population standard deviation of the paired difference between the later and initial levels of FEV1 is 25%.)

3 Fat-free mass (FFM) is an important indicator of physical health. In a recent study, the FFM of 57 pre-pubertal boys was measured by two methods: a reference method that is known to be extremely accurate and a test method that is easier to carry out. The paired difference, test FFM *minus* reference FFM (kg), was determined for each boy. The sample average of these paired differences was 2.72 kg. Using a fixed significance level of 0.01, conduct a formal test of the alternative hypothesis that, on average, the test method gives a different result from the reference method. (You may assume that the population standard deviation of these paired differences is 3 kg.)

## Hypothesis tests and confidence intervals

An engineering firm has been buying shearing pins from one manufacturer. The length of time for which one of these pins can be used before it fractures, known as the **lifetime** of the pin, has been demonstrated to be Normally distributed with mean 100 hours and standard deviation 4 hours. The engineering firm now decides to investigate shearing pins made by another supplier. In tests on 100 of the alternative make of shearing pins, the sample mean lifetime is found to be 99.2 hours. Does it appear that the shearing pins made by this alternative supplier have a different mean lifetime than the pins made by the current supplier?

We assume that the shearing pins made by the alternative supplier have lifetimes that are Normally distributed, with population mean $\mu$ hours and standard deviation 4 hours (the same standard deviation as the pins made by the current supplier).

We wish to investigate the following hypotheses:

$$H_0: \mu = 100 \quad \text{vs} \quad H_1: \mu \neq 100$$

We might choose to proceed by adopting a fixed significance level and finding the critical region for the test. Since $\bar{x} = 99.2$,

$$z = \frac{\bar{x} - \mu_0}{\sigma/\sqrt{n}} = \frac{99.2 - 100}{4/\sqrt{100}} = -2.00$$

If we adopt a significance level $\alpha = 0.05$, then the critical region for this two-tailed test is

$$CR = \{z: z < -z_{0.025} \text{ or } z > z_{0.025}\} = \{z: z < -1.96 \text{ or } z > 1.96\}$$

Alternatively, we might choose to proceed by determining the $p$-value for the test:

$$\begin{aligned} p &= P(Z < -2.00 \mid Z \sim N(0, 1)) + P(Z > 2.00 \mid Z \sim N(0, 1)) \\ &= 2 \times \Phi(-2.00) \\ &= 2 \times \{1 - \Phi(2.00)\} \\ &= 0.0456 \end{aligned}$$

We might even choose to proceed by calculating a confidence interval for $\mu$, and using it to check whether or not the value 100 is plausible for $\mu$. A 95% confidence interval for $\mu$ is

$$\bar{x} - 1.96\frac{\sigma}{\sqrt{n}} \quad \text{to} \quad \bar{x} + 1.96\frac{\sigma}{\sqrt{n}}$$

i.e. $\quad 99.2 - 0.784 \quad$ to $\quad 99.2 + 0.784$

i.e. $\quad 98.416 \quad$ to $\quad 99.984$

We can summarise our conclusions from these three forms of statistical analysis as follows:

| | |
|---|---|
| $z$ is in CR | we reject $H_0: \mu = 100$ at a significance level of 0.05 |
| $p$ is small ($<0.05$) | we reject $H_0: \mu = 100$ |
| 100 is not in the CI | 100 is not a plausible value for $\mu$ |

All three analyses lead us to conclude that the population mean lifetime of the new supplier's shearing pins is significantly different from 100 hours. However, the confidence interval is more informative than the test, since it explicitly indicates the plausible values of the parameter. Here 100 is only narrowly outside the confidence interval. A $p$-value is more informative than the result of a test with a fixed significance level; here $p$ is only just smaller than 0.05.

In general, there is the following close connection between the results of a *two-tailed* test of $H_0: \mu = \mu_0$ and the form of a confidence interval for $\mu$

$p < \alpha$

$H_0$ is rejected at a fixed significance level of $\alpha$

$\mu_0$ is not in a $100(1 - \alpha)\%$ confidence interval for $\mu$

Note that the confidence intervals that were discussed in Chapter 3 are two-sided intervals. That is why they tie up so closely with two-tailed hypothesis tests. It is possible to produce one-sided confidence intervals, which have a similar relationship to one-tailed hypothesis tests, but that is beyond the scope of this course.

In passing, it is worth considering the implications of the above analysis. The alternative supplier's shearing pins are very likely to have a population mean lifetime somewhere between 98.42 hours and 99.98 hours. This population mean is *statistically* significantly different from 100 hours, as we have seen. However, the very small difference might well not be *operationally* significant to the customer, in the sense that it might not matter in practice. It is important always to remember, when interpreting the outcome of an hypothesis test, that statistical significance and operational significance are not the same thing.



## EXERCISE 4.6B

**1** **a** Referring back to question 3 of Exercise 4.2, on the lengths of new-born babies in Glasgow, calculate a 95% confidence interval for the population mean length of **(i)** new-born boys in Glasgow and **(ii)** new-born girls in Glasgow.

**b** Compare your results from part **a** above with your results from question 3 of Exercise 4.3.

**2** **a** Referring back to question 3 of Exercise 4.5B, on the FFM of pre-pubertal boys, calculate a 99% confidence interval for the population mean paired difference between FFM determined by the test and reference methods.

**b** Compare your results from part **a** above with your results from question 3 of Exercise 4.5B.

**3** A line of length 138 mm was drawn on a sheet of paper. 40 second-year statistics students were each asked to guess the length of the line (to the nearest mm). Their sample average guess was 141.25 mm.

**a** Assuming that the population standard deviation of a line estimate made in these circumstances is 27.6 mm, test the hypothesis that second-year statistics students guess the correct length on average. Use a fixed significance level of 0.05.

**b** Calculate a 95% confidence interval for the population mean guess of length.

**c** Compare your results from parts **a** and **b**.

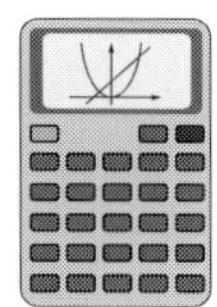

Some calculators will perform hypothesis tests and confidence intervals. Look for menu items such as

Z-Test to test $\mu$ when $\sigma$ is known

ZInterval to construct a confidence interval for $\mu$ when $\sigma$ is known

Find out what your calculator can do.

# STATISTICS IN ACTION – A TEST OF A POPULATION PROPORTION

Cholesterol is a fatty substance that makes up a large portion of the membrane of the cells in your body. If the blood level of cholesterol gets too high, it builds up in the arteries causing inflammation and scarring (arteriosclerosis). If the artery gets completely blocked as a result, oxygen may be prevented from reaching the heart, causing a heart attack. For this reason, doctors recommend keeping your blood cholesterol level low, below 5 to 6 mmol/l.

During the West of Scotland Coronary Prevention Study (WOSCOPS) in the mid 1990s, 80 197 men aged 45 to 64 years were screened for potential heart disease. 22 696 of these men were found to have a blood cholesterol level of 6.5 mmol/l or greater.

**1** Describe the population for this study. Let $p$ be the proportion of people in this population who have a blood cholesterol level of 6.5 mmol/l or greater. Estimate $p$.

**2** Making all necessary assumptions, calculate a 99.9% confidence interval for $p$.

**3** Suppose that a large random sample, of size $n$, was selected from the study population and that $X$ of them were found to have a blood cholesterol level of at least 6.5 mmol/l. Write down the exact distribution of $X$ in terms of $n$ and $p$, and a Normal approximation to this distribution. Hence find a statistic which approximately follows the Standard Normal distribution for large values of $n$.

**4** Now consider testing

$$H_0\colon p = p_0 \quad \text{vs} \quad H_1\colon p > p_0$$

Use the results from part 3 to construct an appropriate test statistic. (*Do not* use a continuity correction.)

**5** Using a fixed significance level of 0.001, investigate the hypothesis that more than $\frac{1}{4}$ of the people in this population have a blood cholesterol level of 6.5 mmol/l or greater.

**6** Interpret carefully the results of all the analyses you have carried out.

# CHAPTER 4 SUMMARY

1 (i) The **null hypothesis** ($H_0$) and the **alternative hypothesis** ($H_1$) are competing statements about a population parameter, such as a population mean or a population proportion.

(ii) In this chapter, we have discussed a test of the population mean where the null hypothesis states that the population mean is *equal* to a known value,

$H_0: \mu = \mu_0$

(iii) The alternative hypothesis can take one of the following forms

$H_1: \mu \neq \mu_0$ **(two-tailed)**
$H_1: \mu < \mu_0$ **(one-tailed)**
$H_1: \mu > \mu_0$ **(one-tailed)**

2 A **hypothesis test** is a statistical procedure that is carried out to assess the evidence for and against $H_0$ and $H_1$ from a sample of data, $X_1, X_2, \ldots, X_n$.

3 (i) A **test statistic** is a function of the sample data, with a probability distribution that is *known* when $H_0$ is true. The test statistic for the hypothesis test considered in this chapter is

$$Z = \frac{\overline{X} - \mu_0}{\sigma/\sqrt{n}} = \frac{\overline{X} - \mu_0}{\text{standard error of } \overline{X}}$$

(ii) Assuming that

- the $X_i$s are all independent,
- the $X_i$s are a sample from a normal population,
- the population standard deviation, $\sigma$, is known,

then $Z \sim N(0, 1)$ when $H_0$ is true.

(iii) The distribution of the test statistic when the null hypothesis is true is called the **null distribution**.

(iv) The value of the test statistic changes from sample to sample. The value obtained from the sample that is actually observed is called the **observed value of the test statistic**, $z$. When $H_0$ is true, we expect to find that $z$ is close to 0, which is the expected value of $Z$ under $H_0$. Observed values $z$ that are close to 0 are consistent with $H_0$, while observed values $z$ that are further away from 0 are more consistent with $H_1$.

4 (i) The ***p*-value** of the hypothesis test is the *conditional* probability of $Z$ being at least as extreme as $z$ *given that* $H_0$ is true. For a *two-tailed* alternative hypothesis,

either $p = P(Z \leq -z \text{ or } Z \geq z \mid H_0)$ when $z$ is non-negative
or $p = P(Z \leq z \text{ or } Z \geq -z \mid H_0)$ when $z$ is negative

(ii) For a *one-tailed* alternative hypothesis,

*either* $p = P(Z \leq z \mid H_0) = \Phi(z)$ when $H_1: \mu < \mu_0$
*or* $p = P(Z \geq z \mid H_0) = 1 - \Phi(z)$ when $H_1: \mu > \mu_0$

(iii) **$H_0$ is rejected in favour of $H_1$** when $p$ is *small*, usually $p < 0.05$ or $p < 0.01$. Otherwise, **$H_0$ is not rejected**.

**5 (i)** Alternatively, a hypothesis test may be carried out at a fixed **significance level**, $\alpha$, where $\alpha$ is a small value, for example $\alpha = 0.05$ or $\alpha = 0.01$. This is equivalent to rejecting $H_0$ when $p \leq \alpha$.

**(ii)** The **critical value** of the test is the value of $z$ that gives a $p$-value exactly equal to $\alpha$, and the **critical region** (CR) is the set of $z$-values that are at least as extreme as the critical value.

**(iii)** Define $z_p$ by $P(Z > z_p | Z \sim N(0, 1)) = p$, where $p$ is any probability between 0 and 1. Then for a *two-tailed* test,

$$CR = \{z: z < -z_{\alpha/2} \text{ or } z > z_{\alpha/2}\}$$

and, for a *one-tailed* test,

*either* $CR = \{z: z < -z_\alpha\}$ when $H_1: \mu < \mu_0$

*or* $CR = \{z: z > z_\alpha\}$ when $H_1: \mu > \mu_0$

**(iv)** $H_0$ is rejected in favour of $H_1$ when $z$ is found to lie in CR. Otherwise, $H_0$ is not rejected.

**6 (i)** When $H_0$ is rejected in favour of $H_1$, we conclude that

*either* $\mu$ is *significantly* different from $\mu_0$

*or* $\mu$ is *significantly* less than $\mu_0$

*or* $\mu$ is *significantly* greater than $\mu_0$

depending on the form of $H_1$.

**(ii)** When $H_0$ is not rejected, we conclude that

*either* $\mu$ is *not significantly* different from $\mu_0$

*or* $\mu$ is *not significantly* less than $\mu_0$

*or* $\mu$ is *not significantly* greater than $\mu_0$

depending on the form of $H_1$.

**7 (i)** This test can be extended to deal with a situation in which $x_i$ and $y_i$ are measurements of the same quantity under two different conditions, and it makes sense to form the **paired difference** $d_i = x_i - y_i$. Often this means that $x_i$ and $y_i$ are measured on the same individual.

**(ii)** Comparison of the population means $\mu_X$ and $\mu_Y$ can be achieved by investigating the population mean of the paired differences, $\mu_D = \mu_X - \mu_Y$. For example, the following hypotheses are equivalent in this context

$H_0: \mu_X = \mu_Y$ and $H_0: \mu_D = 0$

$H_1: \mu_X \neq \mu_Y$ and $H_1: \mu_D \neq 0$

$H_1: \mu_X < \mu_Y$ and $H_1: \mu_D < 0$

$H_1: \mu_X > \mu_Y$ and $H_1: \mu_D > 0$

$H_0$ can be tested against any of these possible alternatives using the $z$ test with the paired differences. The observed value of the appropriate test statistic is

$$z = \frac{\bar{d} - 0}{\sigma_D/\sqrt{n}}$$

**8** The results of a *two-tailed* test of $H_0: \mu = \mu_0$ and the form of a confidence interval for $\mu$ must agree, in the following sense. $p$ is less than $\alpha$, i.e. $H_0$ is rejected at the fixed significance level $\alpha$, if and only if $\mu_0$ does *not* lie in a $100(1 - \alpha)\%$ confidence interval for $\mu$.

# CHAPTER 4 REVIEW EXERCISE

**1** The American physicist Albert Michelson sought to determine the speed of light in a series of famous experiments conducted towards the end of the nineteenth century. The best modern determination of the speed of light is 299 792.5 km/s. Given below are 22 determinations of the speed of light (km/s) made during Michelson's experiments in 1882. (One further, extremely high, observation has been omitted.) To simplify the presentation of the data, the value of 299 000 has been subtracted from each observation.

| 573 | 578 | 599 | 611 | 682 | 696 | 711 | 723 | 748 | 748 | 772 |
|---|---|---|---|---|---|---|---|---|---|---|
| 774 | 778 | 781 | 796 | 796 | 797 | 809 | 816 | 820 | 851 | 883 |

**a** Plot these data on a boxplot. Comment on the assumption that experimental determinations of the speed of light are Normally distributed. Does it appear as though Michelson's experiments, on average, yielded the correct value of the speed of light?

**b** Assuming that experiments like these have a (population) standard deviation of 90 km/s, obtain a 95% confidence interval of the average speed of light as determined by Michelson's experimental technique.

**c** Using a significance level of 0.05, test the null hypothesis that, on average, Michelson's experimental technique would produce the correct value of the speed of light.

**2** **a** UK national growth standards suggest that the length of one-year-old baby boys is 75.80 cm on average, with a standard deviation of 2.53 cm. In a recent study, a random sample of 67 boys who had been born at the Queen Mother's Hospital, Glasgow, was carefully measured at one year of age. The sample average length was found to be 76.06 cm. Investigate the hypothesis that one-year-old baby boys who were born in Glasgow have a different average length from that suggested by the national growth standards.

**b** According to the same growth standards, the length of one-year-old baby girls is 73.99 cm on average, with a standard deviation of 2.41 cm. In the same study, a random sample of 55 one-year-old girls who had been born at the Queen Mother's Hospital were found to have an average length of 74.13 cm. Repeat part **a** for one-year-old baby girls who were born in Glasgow.

**3** A recent study investigated weight loss in patients admitted to a unit for the elderly, for an acute cause (e.g. a broken hip). In the space of one year, 293 patients were admitted, not counting those who were overweight at admission. 151 of these patients were chosen at random to act as controls, while the remaining 142 patients were prescribed a nutritional supplement in addition to the usual hospital diet. All patients were weighed on admission to hospital and then on discharge. Of particular interest was the paired difference, discharge weight *minus* admission weight (kg).

**a** The sample mean difference for the control patients was –0.526 kg (i.e. an average weight *loss* of 0.526 kg). Assuming that the relevant population standard deviation of weight change is 2.75 kg, investigate the study hypothesis that elderly patients lose weight on average during a hospital admission.

**b** The sample mean weight difference for the patients prescribed the nutritional supplement was 0.431 kg (i.e. an average weight *gain* of 0.431 kg). Repeat part **a** for elderly patients who are prescribed a nutritional supplement.

**c** Compare your answers to parts **a** and **b**, and comment.

# Preparation for Unit Assessment

### *Outcome 1 Use conditional probability and the algebra of expectation and variance.*

*(a) Calculate a conditional probability.*

**1** A mail order company employs two stock clerks, Alex and Brenda. Alex processes 40% of the orders and makes a mistake in 1 out of every 100, while Brenda processes the remainder of the orders and makes a mistake in 5 out of every 100. Calculate the probability that an order chosen at random will contain a mistake.

*(b) Apply the laws of expectation and variance in simple cases.*

**2** Two independent random variables, $X$ and $Y$, have means 8 and 6 and standard deviations 3 and 2 respectively. Calculate the mean and variance of the random variables **a** $3X + 2$ **b** $X - Y$.

### *Outcome 2 Use probability distributions in simple situations.*

*(a) Use the Binomial distribution.*

**3** It is thought that 10% of road traffic accidents are due to driver fatigue. What is the probability that exactly two out of five accidents are because of driver fatigue?

*(b) Use the Poisson distribution.*

**4** The number of cars arriving at Dave's Tyre Centre is assumed to follow a Poisson distribution with mean 12 per day. Calculate the probability that on a given day exactly 10 cars arrive.

*(c) Use the Normal distribution.*

**5** In a photographic process, the time (in seconds) to develop a print is distributed N(7.7, 0.25). If 100 prints are produced by this process, how many of them would be expected to take between 7.5 and 7.9 seconds to develop?

### *Outcome 3 Identify sampling methods and estimate population parameters.*

*(a) Identify a given sampling method.*

**6** I wish to interview a sample of Higher Mathematics pupils. To do this, I randomly choose five secondary schools and interview all those studying Higher Mathematics in these schools. Explain which sampling method this illustrates.

*(b) Estimate a population parameter from a sample statistic.*

**7** A random sample of leaves are selected from a large laurel hedge and their lengths measured (mm).

93 98 102 114 120

Estimate the population mean length and variance of such laurel leaves.

## *Outcome 4 Use a z-test on a statistical hypothesis where the significance level is given.*

*(a) State null and alternative hypotheses.*

*(b) State one-tail or two-tail test.*

*(c) Determine the p-value.*

*(d) State and justify an appropriate conclusion.*

**8** The weight of instant coffee which a filling machine places in jars is Normally distributed with mean 103 g and standard deviation 1.6 g.
The filling machine is adjusted and a random sample of 16 jars contains a mean weight of 102.1 g of coffee per jar. Assuming the standard deviation is unchanged, test, at the 5% level, the hypothesis that the adjustment has reduced the mean weight of coffee per jar.

# Preparation for Course Assessment

1 An electronics company receives 75% of its components of a particular type from supplier A and the remainder from supplier B. Of the components supplied by A, 4% are defective while 8% of those supplied by B are defective. Calculate the probability that a randomly selected component, found to be defective, was supplied by A.

2 A diagnostic test for a certain medical condition is such that 90% of the patients who actually have the condition test positive and 80% of the patients who do not have the condition test negative. The remaining patients are wrongly classified. (If a patient tests positive this means the test is indicating they have the condition and if a patient tests negative this means the test is showing that they do not have the condition.) This medical condition affects 10% of the population. If a patient tests positive, what is the probability that he/she actually has the condition?

3 A multiple-choice test consists of ten questions and each question shows four responses of which only one is correct. A particular student answers each question by guessing. Let $X$ denote the number of correct answers achieved by this student.

**a** State the distribution of $X$ and its parameters.

**b** Calculate the probability that the student answers exactly six questions correctly.

4 A fair coin is tossed 100 times. Calculate the approximate probability of obtaining fewer than 45 Heads.

5 The number of mobile phones sold by a retailer is thought to be a Poisson random variable with a mean of 10 per week.

**a** Calculate the probability that during a randomly chosen week the number of mobile phones sold exceeds 10.

**b** State the number of mobile phones which should be in stock at the beginning of a randomly selected week to ensure with at least 95% probability that demand during this period is met from the retailer's stock.

6 Part of an assembly process involves fitting a round peg into a hole. The diameters of these pegs are Normally distributed with mean 10 cm and standard deviation 0.01 cm while the diameters of the holes are distributed Normally with mean 10.05 cm and standard deviation 0.02 cm. Calculate the probability that a randomly chosen peg will not fit into a randomly chosen hole.

7 A random sample of nine eggs laid by a certain breed of hen had the following weights (grams).

40 46 43 40 46 38 45 44 45

You may assume that the weights of these hens' eggs are Normally distributed with a standard deviation of 3.0 g.

**a** Construct a 95% confidence interval for the mean weight of these eggs.

**b** Use this interval to comment on the hypothesis that the mean weight for these eggs is 44 g.

**8** A simple random sample of 65 pupils at a large secondary school were surveyed and 26 of them said they had taken a foreign holiday during the previous year. Construct an approximate 99% confidence interval for the proportion of pupils at this school who have taken a foreign holiday during the past year. Explain what is meant by a 99% confidence interval in this context.

**9** A school has 250 male and 350 female pupils. Explain how you would select a stratified random sample of 60 pupils and how this method of sampling differs from simple random sampling.

**10** An oil company claims that its brand A petrol is better value than a rival brand B because cars travel more miles per gallon when using it. A random sample of eight cars of a particular model were used to test this claim. Each car was filled with 10 gallons of each brand of petrol, driven under controlled conditions round a test track and the total distance travelled on 10 gallons noted. The cars were filled with the two brands of petrol in a random order. The distances travelled were:

| Car | 1 | 2 | 3 | 4 | 5 | 6 | 7 | 8 |
|---|---|---|---|---|---|---|---|---|
| Brand A | 282 | 298 | 274 | 291 | 302 | 285 | 294 | 304 |
| Brand B | 274 | 294 | 275 | 286 | 289 | 273 | 297 | 302 |

Conduct an appropriate statistical test to evaluate the oil company's claim. You may assume that the distances travelled are Normally distributed with $\sigma = 35$ miles.

# Appendices

## Appendix 1 Statistical tables

### Table 1 Binomial cumulative distribution function

The tabulated value is $F(x) = P(X \leq x)$ where $X$ has the Binomial distribution $\text{Bin}(n, p)$. Omitted entries to the left and right of tabulated values are 1.0000 and 0.0000 respectively, to four decimal places.

| | $p =$ | 0.05 | 0.10 | 0.15 | 0.20 | 0.25 | 0.30 | 0.35 | 0.40 | 0.45 | 0.50 |
|---|---|---|---|---|---|---|---|---|---|---|---|
| $n = 4$ | $x = 0$ | 0.8145 | 0.6561 | 0.5220 | 0.4096 | 0.3164 | 0.2401 | 0.1785 | 0.1296 | 0.0915 | 0.0625 |
| | 1 | 0.9860 | 0.9477 | 0.8905 | 0.8192 | 0.7383 | 0.6517 | 0.5630 | 0.4752 | 0.3910 | 0.3125 |
| | 2 | 0.9995 | 0.9963 | 0.9880 | 0.9728 | 0.9492 | 0.9163 | 0.8735 | 0.8208 | 0.7585 | 0.6875 |
| | 3 | | 0.9999 | 0.9995 | 0.9984 | 0.9961 | 0.9919 | 0.9850 | 0.9744 | 0.9590 | 0.9375 |
| $n = 6$ | $x = 0$ | 0.7351 | 0.5314 | 0.3771 | 0.2621 | 0.1780 | 0.1176 | 0.0754 | 0.0467 | 0.0277 | 0.0156 |
| | 1 | 0.9672 | 0.8857 | 0.7765 | 0.6554 | 0.5339 | 0.4202 | 0.3191 | 0.2333 | 0.1636 | 0.1094 |
| | 2 | 0.9978 | 0.9842 | 0.9527 | 0.9011 | 0.8306 | 0.7443 | 0.6471 | 0.5443 | 0.4415 | 0.3438 |
| | 3 | 0.9999 | 0.9987 | 0.9941 | 0.9830 | 0.9624 | 0.9295 | 0.8826 | 0.8208 | 0.7447 | 0.6563 |
| | 4 | | 0.9999 | 0.9996 | 0.9984 | 0.9954 | 0.9891 | 0.9777 | 0.9590 | 0.9308 | 0.8906 |
| | 5 | | | | 0.9999 | 0.9998 | 0.9993 | 0.9982 | 0.9959 | 0.9917 | 0.9844 |
| $n = 8$ | $x = 0$ | 0.6634 | 0.4305 | 0.2725 | 0.1678 | 0.1001 | 0.0576 | 0.0319 | 0.0168 | 0.0084 | 0.0039 |
| | 1 | 0.9428 | 0.8131 | 0.6572 | 0.5033 | 0.3671 | 0.2553 | 0.1691 | 0.1064 | 0.0632 | 0.0352 |
| | 2 | 0.9942 | 0.9619 | 0.8948 | 0.7969 | 0.6785 | 0.5518 | 0.4278 | 0.3154 | 0.2201 | 0.1445 |
| | 3 | 0.9996 | 0.9950 | 0.9786 | 0.9437 | 0.8862 | 0.8059 | 0.7064 | 0.5941 | 0.4770 | 0.3633 |
| | 4 | | 0.9996 | 0.9971 | 0.9896 | 0.9727 | 0.9420 | 0.8939 | 0.8263 | 0.7396 | 0.6367 |
| | 5 | | | 0.9998 | 0.9988 | 0.9958 | 0.9887 | 0.9747 | 0.9502 | 0.9115 | 0.8555 |
| | 6 | | | | 0.9999 | 0.9996 | 0.9987 | 0.9964 | 0.9915 | 0.9819 | 0.9648 |
| | 7 | | | | | | 0.9999 | 0.9998 | 0.9993 | 0.9983 | 0.9961 |
| $n = 10$ | $x = 0$ | 0.5987 | 0.3487 | 0.1969 | 0.1074 | 0.0563 | 0.0282 | 0.0135 | 0.0060 | 0.0025 | 0.0010 |
| | 1 | 0.9139 | 0.7361 | 0.5443 | 0.3758 | 0.2440 | 0.1493 | 0.0860 | 0.0464 | 0.0233 | 0.0107 |
| | 2 | 0.9885 | 0.9298 | 0.8202 | 0.6778 | 0.5256 | 0.3828 | 0.2616 | 0.1673 | 0.0996 | 0.0547 |
| | 3 | 0.9990 | 0.9872 | 0.9500 | 0.8791 | 0.7759 | 0.6496 | 0.5138 | 0.3823 | 0.2660 | 0.1719 |
| | 4 | 0.9999 | 0.9984 | 0.9901 | 0.9672 | 0.9219 | 0.8497 | 0.7515 | 0.6331 | 0.5044 | 0.3770 |
| | 5 | | 0.9999 | 0.9986 | 0.9936 | 0.9803 | 0.9527 | 0.9051 | 0.8338 | 0.7384 | 0.6230 |
| | 6 | | | 0.9999 | 0.9991 | 0.9965 | 0.9894 | 0.9740 | 0.9452 | 0.8980 | 0.8281 |
| | 7 | | | | 0.9999 | 0.9996 | 0.9984 | 0.9952 | 0.9877 | 0.9726 | 0.9453 |
| | 8 | | | | | | 0.9999 | 0.9995 | 0.9983 | 0.9955 | 0.9893 |
| | 9 | | | | | | | | 0.9999 | 0.9997 | 0.9990 |

Table 1 *continued*

| | $p=$ | 0.05 | 0.10 | 0.15 | 0.20 | 0.25 | 0.30 | 0.35 | 0.40 | 0.45 | 0.50 |
|---|---|---|---|---|---|---|---|---|---|---|---|
| $n=12$ | $x=0$ | 0.5404 | 0.2824 | 0.1422 | 0.0687 | 0.0317 | 0.0138 | 0.0057 | 0.0022 | 0.0008 | 0.0002 |
| | 1 | 0.8816 | 0.6590 | 0.4435 | 0.2749 | 0.1584 | 0.0850 | 0.0424 | 0.0196 | 0.0083 | 0.0032 |
| | 2 | 0.9804 | 0.8891 | 0.7358 | 0.5583 | 0.3907 | 0.2528 | 0.1513 | 0.0834 | 0.0421 | 0.0193 |
| | 3 | 0.9978 | 0.9744 | 0.9078 | 0.7946 | 0.6488 | 0.4925 | 0.3467 | 0.2253 | 0.1345 | 0.0730 |
| | 4 | 0.9998 | 0.9957 | 0.9761 | 0.9274 | 0.8424 | 0.7237 | 0.5833 | 0.4382 | 0.3044 | 0.1938 |
| | 5 | | 0.9995 | 0.9954 | 0.9806 | 0.9456 | 0.8822 | 0.7873 | 0.6652 | 0.5269 | 0.3872 |
| | 6 | | 0.9999 | 0.9993 | 0.9961 | 0.9857 | 0.9614 | 0.9154 | 0.8418 | 0.7393 | 0.6128 |
| | 7 | | | 0.9999 | 0.9994 | 0.9972 | 0.9905 | 0.9745 | 0.9427 | 0.8883 | 0.8062 |
| | 8 | | | | 0.9999 | 0.9996 | 0.9983 | 0.9944 | 0.9847 | 0.9644 | 0.9270 |
| | 9 | | | | | | 0.9998 | 0.9992 | 0.9972 | 0.9921 | 0.9807 |
| | 10 | | | | | | | 0.9999 | 0.9997 | 0.9989 | 0.9968 |
| | 11 | | | | | | | | | 0.9999 | 0.9998 |
| $n=14$ | $x=0$ | 0.4877 | 0.2288 | 0.1028 | 0.0440 | 0.0178 | 0.0068 | 0.0024 | 0.0008 | 0.0002 | 0.0001 |
| | 1 | 0.8470 | 0.5846 | 0.3567 | 0.1979 | 0.1010 | 0.0475 | 0.0205 | 0.0081 | 0.0029 | 0.0009 |
| | 2 | 0.9699 | 0.8416 | 0.6479 | 0.4481 | 0.2811 | 0.1608 | 0.0839 | 0.0398 | 0.0170 | 0.0065 |
| | 3 | 0.9958 | 0.9559 | 0.8535 | 0.6982 | 0.5213 | 0.3552 | 0.2205 | 0.1243 | 0.0632 | 0.0287 |
| | 4 | 0.9996 | 0.9908 | 0.9533 | 0.8702 | 0.7415 | 0.5842 | 0.4227 | 0.2793 | 0.1672 | 0.0898 |
| | 5 | | 0.9985 | 0.9885 | 0.9561 | 0.8883 | 0.7805 | 0.6405 | 0.4859 | 0.3373 | 0.2120 |
| | 6 | | 0.9998 | 0.9978 | 0.9884 | 0.9617 | 0.9067 | 0.8164 | 0.6925 | 0.5461 | 0.3953 |
| | 7 | | | 0.9997 | 0.9976 | 0.9897 | 0.9685 | 0.9247 | 0.8499 | 0.7414 | 0.6047 |
| | 8 | | | | 0.9996 | 0.9978 | 0.9917 | 0.9757 | 0.9417 | 0.8811 | 0.7880 |
| | 9 | | | | | 0.9997 | 0.9983 | 0.9940 | 0.9825 | 0.9574 | 0.9102 |
| | 10 | | | | | | 0.9998 | 0.9989 | 0.9961 | 0.9886 | 0.9713 |
| | 11 | | | | | | | 0.9999 | 0.9994 | 0.9978 | 0.9935 |
| | 12 | | | | | | | | 0.9999 | 0.9997 | 0.9991 |
| | 13 | | | | | | | | | | 0.9999 |
| $n=16$ | $x=0$ | 0.4401 | 0.1853 | 0.0743 | 0.0281 | 0.0100 | 0.0033 | 0.0010 | 0.0003 | 0.0001 | |
| | 1 | 0.8108 | 0.5147 | 0.2839 | 0.1407 | 0.0635 | 0.0261 | 0.0098 | 0.0033 | 0.0010 | 0.0003 |
| | 2 | 0.9571 | 0.7892 | 0.5614 | 0.3518 | 0.1971 | 0.0994 | 0.0451 | 0.0183 | 0.0066 | 0.0021 |
| | 3 | 0.9930 | 0.9316 | 0.7899 | 0.5981 | 0.4050 | 0.2459 | 0.1339 | 0.0651 | 0.0281 | 0.0106 |
| | 4 | 0.9991 | 0.9830 | 0.9209 | 0.7982 | 0.6302 | 0.4499 | 0.2892 | 0.1666 | 0.0853 | 0.0384 |
| | 5 | 0.9999 | 0.9967 | 0.9765 | 0.9183 | 0.8103 | 0.6598 | 0.4900 | 0.3288 | 0.1976 | 0.1051 |
| | 6 | | 0.9995 | 0.9944 | 0.9733 | 0.9204 | 0.8247 | 0.6881 | 0.5272 | 0.3660 | 0.2272 |
| | 7 | | 0.9999 | 0.9989 | 0.9930 | 0.9729 | 0.9256 | 0.8406 | 0.7161 | 0.5629 | 0.4018 |
| | 8 | | | 0.9998 | 0.9985 | 0.9925 | 0.9743 | 0.9329 | 0.8577 | 0.7441 | 0.5982 |
| | 9 | | | | 0.9998 | 0.9984 | 0.9929 | 0.9771 | 0.9417 | 0.8759 | 0.7728 |
| | 10 | | | | | 0.9997 | 0.9984 | 0.9938 | 0.9809 | 0.9514 | 0.8949 |
| | 11 | | | | | | 0.9997 | 0.9987 | 0.9951 | 0.9851 | 0.9616 |
| | 12 | | | | | | | 0.9998 | 0.9991 | 0.9965 | 0.9894 |
| | 13 | | | | | | | | 0.9999 | 0.9994 | 0.9979 |
| | 14 | | | | | | | | | 0.9999 | 0.9997 |

Table 1 *continued*

| | $p=$ | 0.05 | 0.10 | 0.15 | 0.20 | 0.25 | 0.30 | 0.35 | 0.40 | 0.45 | 0.50 |
|---|---|---|---|---|---|---|---|---|---|---|---|
| $n=18$ | $x=0$ | 0.3972 | 0.1501 | 0.0536 | 0.0180 | 0.0056 | 0.0016 | 0.0004 | 0.0001 | | |
| | 1 | 0.7735 | 0.4503 | 0.2241 | 0.0991 | 0.0395 | 0.0142 | 0.0046 | 0.0013 | 0.0003 | 0.0001 |
| | 2 | 0.9419 | 0.7338 | 0.4797 | 0.2713 | 0.1353 | 0.0600 | 0.0236 | 0.0082 | 0.0025 | 0.0007 |
| | 3 | 0.9891 | 0.9018 | 0.7202 | 0.5010 | 0.3057 | 0.1646 | 0.0783 | 0.0328 | 0.0120 | 0.0038 |
| | 4 | 0.9985 | 0.9718 | 0.8794 | 0.7164 | 0.5187 | 0.3327 | 0.1886 | 0.0942 | 0.0411 | 0.0154 |
| | 5 | 0.9998 | 0.9936 | 0.9581 | 0.8671 | 0.7175 | 0.5344 | 0.3550 | 0.2088 | 0.1077 | 0.0481 |
| | 6 | | 0.9988 | 0.9882 | 0.9487 | 0.8610 | 0.7217 | 0.5491 | 0.3743 | 0.2258 | 0.1189 |
| | 7 | | 0.9998 | 0.9973 | 0.9837 | 0.9431 | 0.8593 | 0.7283 | 0.5634 | 0.3915 | 0.2403 |
| | 8 | | | 0.9995 | 0.9957 | 0.9807 | 0.9404 | 0.8609 | 0.7368 | 0.5778 | 0.4073 |
| | 9 | | | 0.9999 | 0.9991 | 0.9946 | 0.9790 | 0.9403 | 0.8653 | 0.7473 | 0.5927 |
| | 10 | | | | 0.9998 | 0.9988 | 0.9939 | 0.9788 | 0.9424 | 0.8720 | 0.7597 |
| | 11 | | | | | 0.9998 | 0.9986 | 0.9938 | 0.9797 | 0.9463 | 0.8811 |
| | 12 | | | | | | 0.9997 | 0.9986 | 0.9942 | 0.9817 | 0.9519 |
| | 13 | | | | | | | 0.9997 | 0.9987 | 0.9951 | 0.9846 |
| | 14 | | | | | | | | 0.9998 | 0.9990 | 0.9962 |
| | 15 | | | | | | | | | 0.9999 | 0.9993 |
| | 16 | | | | | | | | | | 0.9999 |
| $n=20$ | $x=0$ | 0.3585 | 0.1216 | 0.0388 | 0.0115 | 0.0032 | 0.0008 | 0.0002 | | | |
| | 1 | 0.7358 | 0.3917 | 0.1756 | 0.0692 | 0.0243 | 0.0076 | 0.0021 | 0.0005 | 0.0001 | |
| | 2 | 0.9245 | 0.6769 | 0.4049 | 0.2061 | 0.0913 | 0.0355 | 0.0121 | 0.0036 | 0.0009 | 0.0002 |
| | 3 | 0.9841 | 0.8670 | 0.6477 | 0.4114 | 0.2252 | 0.1071 | 0.0444 | 0.0160 | 0.0049 | 0.0013 |
| | 4 | 0.9974 | 0.9568 | 0.8298 | 0.6296 | 0.4148 | 0.2375 | 0.1182 | 0.0510 | 0.0189 | 0.0059 |
| | 5 | 0.9997 | 0.9887 | 0.9327 | 0.8042 | 0.6172 | 0.4164 | 0.2454 | 0.1256 | 0.0553 | 0.0207 |
| | 6 | | 0.9976 | 0.9781 | 0.9133 | 0.7858 | 0.6080 | 0.4166 | 0.2500 | 0.1299 | 0.0577 |
| | 7 | | 0.9996 | 0.9941 | 0.9679 | 0.8982 | 0.7723 | 0.6010 | 0.4159 | 0.2520 | 0.1316 |
| | 8 | | 0.9999 | 0.9987 | 0.9900 | 0.9591 | 0.8867 | 0.7624 | 0.5956 | 0.4143 | 0.2517 |
| | 9 | | | 0.9998 | 0.9974 | 0.9861 | 0.9520 | 0.8782 | 0.7553 | 0.5914 | 0.4119 |
| | 10 | | | | 0.9994 | 0.9961 | 0.9829 | 0.9468 | 0.8725 | 0.7507 | 0.5881 |
| | 11 | | | | 0.9999 | 0.9991 | 0.9949 | 0.9804 | 0.9435 | 0.8692 | 0.7483 |
| | 12 | | | | | 0.9998 | 0.9987 | 0.9940 | 0.9790 | 0.9420 | 0.8684 |
| | 13 | | | | | | 0.9997 | 0.9985 | 0.9935 | 0.9786 | 0.9423 |
| | 14 | | | | | | | 0.9997 | 0.9984 | 0.9936 | 0.9793 |
| | 15 | | | | | | | | 0.9997 | 0.9985 | 0.9941 |
| | 16 | | | | | | | | | 0.9997 | 0.9987 |
| | 17 | | | | | | | | | | 0.9998 |

### *Table 2 Poisson cumulative distribution function*

The tabulated value is $F(x) = P(X \leq x)$ where $X$ has the Poisson distribution Poi($\mu$). Omitted entries to the left and right of tabulated values are 1.0000 and 0.0000 respectively, to four decimal places.

| $\mu =$ | 0.5 | 1.0 | 1.5 | 2.0 | 2.5 | 3.0 | 3.5 | 4.0 | 4.5 | 5.0 |
|---|---|---|---|---|---|---|---|---|---|---|
| $x = 0$ | 0.6065 | 0.3679 | 0.2231 | 0.1353 | 0.0821 | 0.0498 | 0.0302 | 0.0183 | 0.0111 | 0.0067 |
| 1 | 0.9098 | 0.7358 | 0.5578 | 0.4060 | 0.2873 | 0.1991 | 0.1359 | 0.0916 | 0.0611 | 0.0404 |
| 2 | 0.9856 | 0.9197 | 0.8088 | 0.6767 | 0.5438 | 0.4232 | 0.3208 | 0.2381 | 0.1736 | 0.1247 |
| 3 | 0.9982 | 0.9810 | 0.9344 | 0.8571 | 0.7576 | 0.6472 | 0.5366 | 0.4335 | 0.3423 | 0.2650 |
| 4 | 0.9998 | 0.9963 | 0.9814 | 0.9473 | 0.8912 | 0.8153 | 0.7254 | 0.6288 | 0.5321 | 0.4405 |
| 5 | | 0.9994 | 0.9955 | 0.9834 | 0.9580 | 0.9161 | 0.8576 | 0.7851 | 0.7029 | 0.6160 |
| 6 | | 0.9999 | 0.9991 | 0.9955 | 0.9858 | 0.9665 | 0.9347 | 0.8893 | 0.8311 | 0.7622 |
| 7 | | | 0.9998 | 0.9989 | 0.9958 | 0.9881 | 0.9733 | 0.9489 | 0.9134 | 0.8666 |
| 8 | | | | 0.9998 | 0.9989 | 0.9962 | 0.9901 | 0.9786 | 0.9597 | 0.9319 |
| 9 | | | | | 0.9997 | 0.9989 | 0.9967 | 0.9919 | 0.9829 | 0.9682 |
| 10 | | | | | 0.9999 | 0.9997 | 0.9990 | 0.9972 | 0.9933 | 0.9863 |
| 11 | | | | | | 0.9999 | 0.9997 | 0.9991 | 0.9976 | 0.9945 |
| 12 | | | | | | | 0.9999 | 0.9997 | 0.9992 | 0.9980 |
| 13 | | | | | | | | 0.9999 | 0.9997 | 0.9993 |
| 14 | | | | | | | | | 0.9999 | 0.9998 |
| 15 | | | | | | | | | | 0.9999 |

| $\mu =$ | 5.5 | 6.0 | 6.5 | 7.0 | 7.5 | 8.0 | 8.5 | 9.0 | 9.5 | 10.0 |
|---|---|---|---|---|---|---|---|---|---|---|
| $x = 0$ | 0.0041 | 0.0025 | 0.0015 | 0.0009 | 0.0006 | 0.0003 | 0.0002 | 0.0001 | 0.0001 | |
| 1 | 0.0266 | 0.0174 | 0.0113 | 0.0073 | 0.0047 | 0.0030 | 0.0019 | 0.0012 | 0.0008 | 0.0005 |
| 2 | 0.0884 | 0.0620 | 0.0430 | 0.0296 | 0.0203 | 0.0138 | 0.0093 | 0.0062 | 0.0042 | 0.0028 |
| 3 | 0.2017 | 0.1512 | 0.1118 | 0.0818 | 0.0591 | 0.0424 | 0.0301 | 0.0212 | 0.0149 | 0.0103 |
| 4 | 0.3575 | 0.2851 | 0.2237 | 0.1730 | 0.1321 | 0.0996 | 0.0744 | 0.0550 | 0.0403 | 0.0293 |
| 5 | 0.5289 | 0.4457 | 0.3690 | 0.3007 | 0.2414 | 0.1912 | 0.1496 | 0.1157 | 0.0885 | 0.0671 |
| 6 | 0.6860 | 0.6063 | 0.5265 | 0.4497 | 0.3782 | 0.3134 | 0.2562 | 0.2068 | 0.1649 | 0.1301 |
| 7 | 0.8095 | 0.7440 | 0.6728 | 0.5987 | 0.5246 | 0.4530 | 0.3856 | 0.3239 | 0.2687 | 0.2202 |
| 8 | 0.8944 | 0.8472 | 0.7916 | 0.7291 | 0.6620 | 0.5925 | 0.5231 | 0.4557 | 0.3918 | 0.3328 |
| 9 | 0.9462 | 0.9161 | 0.8774 | 0.8305 | 0.7764 | 0.7166 | 0.6530 | 0.5874 | 0.5218 | 0.4579 |
| 10 | 0.9747 | 0.9574 | 0.9332 | 0.9015 | 0.8622 | 0.8159 | 0.7634 | 0.7060 | 0.6453 | 0.5830 |
| 11 | 0.9890 | 0.9799 | 0.9661 | 0.9467 | 0.9208 | 0.8881 | 0.8487 | 0.8030 | 0.7520 | 0.6968 |
| 12 | 0.9955 | 0.9912 | 0.9840 | 0.9730 | 0.9573 | 0.9362 | 0.9091 | 0.8758 | 0.8364 | 0.7916 |
| 13 | 0.9983 | 0.9964 | 0.9929 | 0.9872 | 0.9784 | 0.9658 | 0.9486 | 0.9261 | 0.8981 | 0.8645 |
| 14 | 0.9994 | 0.9986 | 0.9970 | 0.9943 | 0.9897 | 0.9827 | 0.9726 | 0.9585 | 0.9400 | 0.9165 |
| 15 | 0.9998 | 0.9995 | 0.9988 | 0.9976 | 0.9954 | 0.9918 | 0.9862 | 0.9780 | 0.9665 | 0.9513 |
| 16 | 0.9999 | 0.9998 | 0.9996 | 0.9990 | 0.9980 | 0.9963 | 0.9934 | 0.9889 | 0.9823 | 0.9730 |
| 17 | | 0.9999 | 0.9998 | 0.9996 | 0.9992 | 0.9984 | 0.9970 | 0.9947 | 0.9911 | 0.9857 |
| 18 | | | 0.9999 | 0.9999 | 0.9997 | 0.9993 | 0.9987 | 0.9976 | 0.9957 | 0.9928 |
| 19 | | | | | 0.9999 | 0.9997 | 0.9995 | 0.9989 | 0.9980 | 0.9965 |
| 20 | | | | | | 0.9999 | 0.9998 | 0.9996 | 0.9991 | 0.9984 |
| 21 | | | | | | | 0.9999 | 0.9998 | 0.9996 | 0.9993 |
| 22 | | | | | | | | 0.9999 | 0.9999 | 0.9997 |
| 23 | | | | | | | | | 0.9999 | 0.9999 |

### *Table 3 Standard Normal cumulative distribution function*

The tabulated value is $\Phi(z) = P(Z \leq z)$ where $Z$ has Standard Normal distribution N(0, 1).

| z | .00 | .01 | .02 | .03 | .04 | .05 | .06 | .07 | .08 | .09 |
|---|---|---|---|---|---|---|---|---|---|---|
| 0.0 | 0.5000 | 0.5040 | 0.5080 | 0.5120 | 0.5160 | 0.5199 | 0.5239 | 0.5279 | 0.5319 | 0.5359 |
| 0.1 | 0.5398 | 0.5438 | 0.5478 | 0.5517 | 0.5557 | 0.5596 | 0.5636 | 0.5675 | 0.5714 | 0.5753 |
| 0.2 | 0.5793 | 0.5832 | 0.5871 | 0.5910 | 0.5948 | 0.5987 | 0.6026 | 0.6064 | 0.6103 | 0.6141 |
| 0.3 | 0.6179 | 0.6217 | 0.6255 | 0.6293 | 0.6331 | 0.6368 | 0.6406 | 0.6443 | 0.6480 | 0.6517 |
| 0.4 | 0.6554 | 0.6591 | 0.6628 | 0.6664 | 0.6700 | 0.6736 | 0.6772 | 0.6808 | 0.6844 | 0.6879 |
| 0.5 | 0.6915 | 0.6950 | 0.6985 | 0.7019 | 0.7054 | 0.7088 | 0.7123 | 0.7157 | 0.7190 | 0.7224 |
| 0.6 | 0.7257 | 0.7291 | 0.7324 | 0.7357 | 0.7389 | 0.7422 | 0.7454 | 0.7486 | 0.7517 | 0.7549 |
| 0.7 | 0.7580 | 0.7611 | 0.7642 | 0.7673 | 0.7704 | 0.7734 | 0.7764 | 0.7794 | 0.7823 | 0.7852 |
| 0.8 | 0.7881 | 0.7910 | 0.7939 | 0.7967 | 0.7995 | 0.8023 | 0.8051 | 0.8078 | 0.8106 | 0.8133 |
| 0.9 | 0.8159 | 0.8186 | 0.8212 | 0.8238 | 0.8264 | 0.8289 | 0.8315 | 0.8340 | 0.8365 | 0.8389 |
| 1.0 | 0.8413 | 0.8438 | 0.8461 | 0.8485 | 0.8508 | 0.8531 | 0.8554 | 0.8577 | 0.8599 | 0.8621 |
| 1.1 | 0.8643 | 0.8665 | 0.8686 | 0.8708 | 0.8729 | 0.8749 | 0.8770 | 0.8790 | 0.8810 | 0.8830 |
| 1.2 | 0.8849 | 0.8869 | 0.8888 | 0.8907 | 0.8925 | 0.8944 | 0.8962 | 0.8980 | 0.8997 | 0.9015 |
| 1.3 | 0.9032 | 0.9049 | 0.9066 | 0.9082 | 0.9099 | 0.9115 | 0.9131 | 0.9147 | 0.9162 | 0.9177 |
| 1.4 | 0.9192 | 0.9207 | 0.9222 | 0.9236 | 0.9251 | 0.9265 | 0.9279 | 0.9292 | 0.9306 | 0.9319 |
| 1.5 | 0.9332 | 0.9345 | 0.9357 | 0.9370 | 0.9382 | 0.9394 | 0.9406 | 0.9418 | 0.9429 | 0.9441 |
| 1.6 | 0.9452 | 0.9463 | 0.9474 | 0.9484 | 0.9495 | 0.9505 | 0.9515 | 0.9525 | 0.9535 | 0.9545 |
| 1.7 | 0.9554 | 0.9564 | 0.9573 | 0.9582 | 0.9591 | 0.9599 | 0.9608 | 0.9616 | 0.9625 | 0.9633 |
| 1.8 | 0.9641 | 0.9649 | 0.9656 | 0.9664 | 0.9671 | 0.9678 | 0.9686 | 0.9693 | 0.9699 | 0.9706 |
| 1.9 | 0.9713 | 0.9719 | 0.9726 | 0.9732 | 0.9738 | 0.9744 | 0.9750 | 0.9756 | 0.9761 | 0.9767 |
| 2.0 | 0.9772 | 0.9778 | 0.9783 | 0.9788 | 0.9793 | 0.9798 | 0.9803 | 0.9808 | 0.9812 | 0.9817 |
| 2.1 | 0.9821 | 0.9826 | 0.9830 | 0.9834 | 0.9838 | 0.9842 | 0.9846 | 0.9850 | 0.9854 | 0.9857 |
| 2.2 | 0.9861 | 0.9864 | 0.9868 | 0.9871 | 0.9875 | 0.9878 | 0.9881 | 0.9884 | 0.9887 | 0.9890 |
| 2.3 | 0.9893 | 0.9896 | 0.9898 | 0.9901 | 0.9904 | 0.9906 | 0.9909 | 0.9911 | 0.9913 | 0.9916 |
| 2.4 | 0.9918 | 0.9920 | 0.9922 | 0.9925 | 0.9927 | 0.9929 | 0.9931 | 0.9932 | 0.9934 | 0.9936 |
| 2.5 | 0.9938 | 0.9940 | 0.9941 | 0.9943 | 0.9945 | 0.9946 | 0.9948 | 0.9949 | 0.9951 | 0.9952 |
| 2.6 | 0.9953 | 0.9955 | 0.9956 | 0.9957 | 0.9959 | 0.9960 | 0.9961 | 0.9962 | 0.9963 | 0.9964 |
| 2.7 | 0.9965 | 0.9966 | 0.9967 | 0.9968 | 0.9969 | 0.9970 | 0.9971 | 0.9972 | 0.9973 | 0.9974 |
| 2.8 | 0.9974 | 0.9975 | 0.9976 | 0.9977 | 0.9977 | 0.9978 | 0.9979 | 0.9979 | 0.9980 | 0.9981 |
| 2.9 | 0.9981 | 0.9982 | 0.9982 | 0.9983 | 0.9984 | 0.9984 | 0.9985 | 0.9985 | 0.9986 | 0.9986 |
| 3.0 | 0.9987 | 0.9987 | 0.9987 | 0.9988 | 0.9988 | 0.9989 | 0.9989 | 0.9989 | 0.9990 | 0.9990 |
| 3.1 | 0.9990 | 0.9991 | 0.9991 | 0.9991 | 0.9992 | 0.9992 | 0.9992 | 0.9992 | 0.9993 | 0.9993 |
| 3.2 | 0.9993 | 0.9993 | 0.9994 | 0.9994 | 0.9994 | 0.9994 | 0.9994 | 0.9995 | 0.9995 | 0.9995 |
| 3.3 | 0.9995 | 0.9995 | 0.9995 | 0.9996 | 0.9996 | 0.9996 | 0.9996 | 0.9996 | 0.9996 | 0.9997 |
| 3.4 | 0.9997 | 0.9997 | 0.9997 | 0.9997 | 0.9997 | 0.9997 | 0.9997 | 0.9997 | 0.9997 | 0.9998 |
| 3.5 | 0.9998 | 0.9998 | 0.9998 | 0.9998 | 0.9998 | 0.9998 | 0.9998 | 0.9998 | 0.9998 | 0.9998 |
| 3.6 | 0.9998 | 0.9998 | 0.9999 | 0.9999 | 0.9999 | 0.9999 | 0.9999 | 0.9999 | 0.9999 | 0.9999 |

For $x > 3.69$ use $\Phi(z) = 1.0000$.

### *Table 4 Percentage points of the Standard Normal distribution*

The entries in the table are such that for the Standard Normal distribution $P(Z > z_p) = p$.

| $p$ | $z_p$ |
|---|---|
| 0.500 | 0.00 |
| 0.250 | 0.67 |
| 0.100 | 1.28 |
| 0.050 | 1.64 |
| 0.025 | 1.96 |
| 0.010 | 2.33 |
| 0.005 | 2.58 |
| 0.001 | 3.09 |
| 0.0005 | 3.29 |

## *Appendix 3 Data on family size*

These data give the number of children in families of pupils at a particular school.
A '4' should be interpreted as '4 or more'.
Each column records the data for one of the 24 register classes.
The first row of the heading gives the number of the register class.
The second row of the heading gives the number of pupils in each register class.

| **1** | **2** | **3** | **4** | **5** | **6** | **7** | **8** | **9** | **10** | **11** | **12** | **13** | **14** | **15** | **16** | **17** | **18** | **19** | **20** | **21** | **22** | **23** | **24** |
|---|---|---|---|---|---|---|---|---|---|---|---|---|---|---|---|---|---|---|---|---|---|---|---|
| 33 | 33 | 33 | 33 | 32 | 33 | 32 | 33 | 33 | 33 | 32 | 32 | 32 | 33 | 31 | 32 | 28 | 25 | 22 | 25 | 18 | 22 | 19 | 21 |
| 2 | 3 | 2 | 2 | 2 | 2 | 2 | 2 | 2 | 2 | 2 | 4 | 2 | 3 | 2 | 2 | 3 | 3 | 2 | 2 | 2 | 2 | 1 | 1 |
| 2 | 2 | 2 | 2 | 3 | 3 | 4 | 3 | 1 | 3 | 2 | 1 | 3 | 2 | 1 | 2 | 2 | 2 | 2 | 1 | 3 | 3 | 2 | 3 |
| 1 | 2 | 3 | 2 | 2 | 2 | 2 | 3 | 2 | 2 | 2 | 3 | 2 | 2 | 3 | 3 | 3 | 2 | 2 | 4 | 2 | 1 | 2 | 1 |
| 2 | 2 | 3 | 2 | 1 | 2 | 2 | 3 | 3 | 3 | 2 | 2 | 3 | 2 | 2 | 3 | 1 | 3 | 2 | 3 | 3 | 2 | 2 | 2 |
| 2 | 3 | 3 | 1 | 2 | 2 | 2 | 1 | 2 | 2 | 2 | 2 | 2 | 2 | 3 | 2 | 3 | 2 | 2 | 2 | 2 | 3 | 1 | 3 |
| 3 | 2 | 2 | 1 | 1 | 4 | 2 | 3 | 3 | 2 | 2 | 2 | 2 | 4 | 2 | 4 | 2 | 2 | 3 | 2 | 2 | 4 | 3 | 2 |
| 2 | 1 | 2 | 3 | 2 | 1 | 2 | 2 | 2 | 2 | 1 | 4 | 4 | 3 | 4 | 3 | 3 | 4 | 1 | 1 | 4 | 2 | 2 | 2 |
| 2 | 2 | 2 | 2 | 3 | 3 | 2 | 1 | 2 | 2 | 3 | 2 | 3 | 2 | 2 | 2 | 2 | 2 | 1 | 2 | 2 | 4 | 2 | 2 |
| 3 | 3 | 2 | 4 | 1 | 1 | 3 | 2 | 1 | 1 | 4 | 2 | 2 | 3 | 2 | 2 | 4 | 2 | 3 | 3 | 2 | 2 | 3 | 2 |
| 2 | 2 | 3 | 2 | 2 | 3 | 2 | 3 | 3 | 1 | 2 | 2 | 4 | 2 | 4 | 2 | 2 | 2 | 2 | 2 | 3 | 3 | 2 | 2 |
| 2 | 1 | 2 | 3 | 3 | 4 | 2 | 2 | 1 | 1 | 3 | 2 | 2 | 2 | 2 | 3 | 3 | 4 | 4 | 3 | 2 | 1 | 2 | 2 |
| 2 | 3 | 2 | 3 | 3 | 2 | 2 | 2 | 3 | 2 | 2 | 3 | 2 | 2 | 2 | 2 | 4 | 2 | 3 | 2 | 2 | 3 | 2 | 2 |
| 2 | 2 | 2 | 2 | 3 | 2 | 3 | 2 | 2 | 2 | 2 | 2 | 2 | 3 | 2 | 2 | 4 | 4 | 2 | 1 | 3 | 1 | 3 | 2 |
| 2 | 1 | 2 | 2 | 2 | 2 | 2 | 1 | 3 | 4 | 3 | 3 | 3 | 2 | 2 | 1 | 2 | 1 | 2 | 3 | 2 | 2 | 2 | 3 |
| 1 | 3 | 2 | 2 | 2 | 3 | 2 | 2 | 2 | 4 | 2 | 3 | 3 | 2 | 1 | 3 | 3 | 2 | 3 | 3 | 2 | 3 | 4 | 4 |
| 3 | 3 | 2 | 3 | 2 | 3 | 3 | 3 | 4 | 3 | 2 | 2 | 3 | 1 | 2 | 2 | 2 | 2 | 2 | 2 | 1 | 2 | 3 | 2 |
| 2 | 2 | 1 | 2 | 3 | 2 | 1 | 2 | 3 | 3 | 3 | 3 | 2 | 2 | 3 | 2 | 2 | 3 | 3 | 2 | 2 | 2 | 2 | 2 |
| 2 | 2 | 4 | 3 | 2 | 3 | 2 | 3 | 2 | 2 | 4 | 3 | 3 | 4 | 3 | 2 | 2 | 2 | 2 | 4 | 2 | 2 | 4 | 2 |
| 2 | 3 | 2 | 2 | 2 | 2 | 2 | 2 | 3 | 4 | 1 | 2 | 2 | 1 | 1 | 2 | 2 | 3 | 2 | 1 |  | 2 | 3 | 2 |
| 2 | 3 | 3 | 4 | 3 | 3 | 2 | 2 | 3 | 3 | 2 | 2 | 2 | 1 | 3 | 2 | 1 | 3 | 1 | 2 |  | 2 |  | 2 |
| 1 | 3 | 2 | 3 | 2 | 1 | 2 | 2 | 2 | 2 | 2 | 2 | 2 | 3 | 2 | 2 | 2 | 2 | 2 | 1 |  | 2 |  | 2 |
| 2 | 4 | 2 | 3 | 2 | 2 | 2 | 2 | 1 | 2 | 2 | 2 | 3 | 2 | 4 | 2 | 2 | 4 | 3 | 3 |  | 3 |  |  |
| 2 | 3 | 2 | 3 | 2 | 2 | 2 | 2 | 2 | 3 | 4 | 3 | 2 | 2 | 1 | 1 | 3 | 3 |  | 1 |  |  |  |  |
| 3 | 2 | 3 | 2 | 3 | 3 | 3 | 4 | 3 | 3 | 2 | 3 | 1 | 3 | 2 | 2 | 1 | 2 |  | 2 |  |  |  |  |
| 3 | 3 | 4 | 3 | 1 | 2 | 2 | 1 | 3 | 1 | 2 | 2 | 2 | 3 | 2 | 1 | 2 | 2 |  | 2 |  |  |  |  |
| 3 | 1 | 3 | 3 | 3 | 2 | 2 | 2 | 1 | 3 | 3 | 3 | 2 | 2 | 3 | 4 | 2 |  |  |  |  |  |  |  |
| 2 | 3 | 1 | 4 | 3 | 2 | 1 | 2 | 3 | 4 | 1 | 4 | 2 | 2 | 3 | 2 | 4 |  |  |  |  |  |  |  |
| 4 | 2 | 1 | 2 | 4 | 2 | 2 | 4 | 2 | 1 | 2 | 2 | 2 | 2 | 2 | 3 | 2 |  |  |  |  |  |  |  |
| 2 | 2 | 2 | 3 | 2 | 2 | 2 | 3 | 2 | 2 | 3 | 4 | 4 | 2 | 2 | 2 |  |  |  |  |  |  |  |  |
| 2 | 2 | 2 | 2 | 1 | 1 | 3 | 2 | 2 | 3 | 2 | 2 | 4 | 2 | 2 | 4 |  |  |  |  |  |  |  |  |
| 3 | 2 | 2 | 3 | 2 | 4 | 2 | 3 | 1 | 2 | 1 | 4 | 2 | 3 | 3 | 2 |  |  |  |  |  |  |  |  |
| 2 | 4 | 3 | 4 | 2 | 3 | 2 | 2 | 2 | 3 | 2 | 1 | 3 | 2 |  | 3 |  |  |  |  |  |  |  |  |
| 4 | 1 | 2 | 2 |  | 3 |  | 4 | 4 | 3 |  |  |  | 1 |  |  |  |  |  |  |  |  |  |  |

# *Appendix 4 Data on wheat yields*

| | Plot | | | | | | | | | |
|---|---|---|---|---|---|---|---|---|---|---|
| Row | 0 | 1 | 2 | 3 | 4 | 5 | 6 | 7 | 8 | 9 |
| 00 | 635 | 585 | 465 | 430 | 615 | 550 | 515 | 369 | 493 | 635 |
| 01 | 655 | 530 | 455 | 465 | 560 | 550 | 460 | 455 | 503 | 519 |
| 02 | 775 | 615 | 545 | 530 | 685 | 595 | 510 | 507 | 561 | 581 |
| 03 | 705 | 555 | 440 | 455 | 630 | 555 | 425 | 476 | 648 | 532 |
| 04 | 655 | 495 | 435 | 485 | 585 | 475 | 405 | 422 | 516 | 458 |
| 05 | 635 | 495 | 445 | 455 | 620 | 505 | 465 | 419 | 591 | 545 |
| 06 | 630 | 555 | 455 | 510 | 575 | 530 | 470 | 427 | 545 | 562 |
| 07 | 675 | 610 | 540 | 530 | 675 | 580 | 460 | 513 | 599 | 595 |
| 08 | 645 | 490 | 445 | 440 | 585 | 490 | 420 | 460 | 542 | 474 |
| 09 | 540 | 495 | 415 | 445 | 530 | 440 | 430 | 455 | 485 | 461 |
| 10 | 670 | 560 | 485 | 445 | 590 | 555 | 450 | 516 | 570 | 553 |
| 11 | 630 | 540 | 465 | 460 | 560 | 465 | 435 | 466 | 454 | 458 |
| 12 | 535 | 525 | 455 | 415 | 510 | 460 | 420 | 393 | 482 | 426 |
| 13 | 495 | 540 | 480 | 395 | 555 | 440 | 450 | 437 | 541 | 525 |
| 14 | 645 | 595 | 515 | 470 | 550 | 515 | 445 | 508 | 508 | 517 |
| 15 | 730 | 710 | 565 | 465 | 640 | 565 | 550 | 494 | 591 | 594 |
| 16 | 635 | 595 | 485 | 435 | 530 | 490 | 470 | 465 | 521 | 454 |
| 17 | 600 | 530 | 455 | 425 | 545 | 395 | 455 | 482 | 484 | 505 |
| 18 | 685 | 685 | 520 | 480 | 525 | 550 | 465 | 497 | 596 | 563 |
| 19 | 605 | 530 | 440 | 450 | 500 | 470 | 445 | 510 | 531 | 478 |
| 20 | 525 | 610 | 515 | 395 | 510 | 455 | 530 | 485 | 554 | 448 |
| 21 | 625 | 600 | 440 | 435 | 510 | 515 | 525 | 500 | 522 | 496 |
| 22 | 580 | 630 | 485 | 480 | 610 | 525 | 515 | 574 | 525 | 458 |
| 23 | 650 | 695 | 550 | 515 | 685 | 570 | 525 | 613 | 560 | 532 |
| 24 | 610 | 545 | 465 | 460 | 555 | 500 | 575 | 438 | 477 | 480 |
| 25 | 555 | 580 | 505 | 465 | 540 | 425 | 490 | 467 | 479 | 444 |
| 26 | 635 | 660 | 550 | 525 | 645 | 515 | 520 | 477 | 510 | 465 |
| 27 | 660 | 595 | 495 | 465 | 595 | 450 | 505 | 417 | 489 | 399 |
| 28 | 530 | 600 | 460 | 410 | 500 | 460 | 505 | 459 | 431 | 438 |
| 29 | 555 | 570 | 450 | 425 | 490 | 450 | 495 | 426 | 443 | 449 |
| 30 | 680 | 685 | 505 | 535 | 600 | 495 | 480 | 474 | 401 | 495 |
| 31 | 715 | 740 | 590 | 585 | 585 | 535 | 560 | 518 | 513 | 541 |
| 32 | 635 | 620 | 495 | 510 | 585 | 520 | 475 | 444 | 405 | 509 |
| 33 | 580 | 505 | 490 | 505 | 555 | 450 | 470 | 468 | 442 | 474 |
| 34 | 645 | 630 | 600 | 565 | 645 | 530 | 550 | 454 | 474 | 592 |
| 35 | 615 | 640 | 505 | 550 | 615 | 535 | 500 | 465 | 456 | 513 |
| 36 | 625 | 685 | 570 | 575 | 685 | 540 | 475 | 534 | 456 | 467 |
| 37 | 700 | 770 | 540 | 515 | 685 | 530 | 495 | 508 | 454 | 422 |
| 38 | 660 | 705 | 570 | 545 | 645 | 500 | 540 | 494 | 500 | 532 |
| 39 | 730 | 825 | 665 | 605 | 665 | 520 | 550 | 584 | 546 | 537 |
| 40 | 640 | 710 | 500 | 535 | 575 | 495 | 455 | 512 | 497 | 465 |
| 41 | 590 | 675 | 575 | 515 | 570 | 490 | 470 | 460 | 427 | 472 |
| 42 | 790 | 855 | 625 | 575 | 655 | 530 | 540 | 575 | 502 | 503 |
| 43 | 705 | 770 | 575 | 560 | 565 | 510 | 435 | 486 | 470 | 502 |
| 44 | 620 | 685 | 620 | 550 | 520 | 435 | 460 | 429 | 485 | 534 |
| 45 | 605 | 680 | 570 | 560 | 560 | 445 | 465 | 461 | 463 | 533 |
| 46 | 760 | 790 | 645 | 605 | 595 | 560 | 485 | 466 | 487 | 476 |
| 47 | 795 | 850 | 700 | 655 | 635 | 540 | 470 | 521 | 582 | 584 |
| 48 | 685 | 720 | 610 | 520 | 550 | 460 | 440 | 425 | 485 | 462 |
| 49 | 695 | 665 | 630 | 510 | 575 | 445 | 430 | 433 | 423 | 450 |

Data on wheat yields *continued*

| | Plot | | | | | | | | | |
|---|---|---|---|---|---|---|---|---|---|---|
| Row | 0 | 1 | 2 | 3 | 4 | 5 | 6 | 7 | 8 | 9 |
| 50 | 840 | 875 | 640 | 575 | 625 | 570 | 490 | 514 | 539 | 486 |
| 51 | 780 | 790 | 575 | 520 | 600 | 545 | 470 | 455 | 486 | 488 |
| 52 | 745 | 840 | 710 | 640 | 610 | 500 | 530 | 498 | 479 | 434 |
| 53 | 735 | 840 | 640 | 610 | 585 | 560 | 450 | 469 | 562 | 450 |
| 54 | 880 | 795 | 660 | 580 | 615 | 585 | 460 | 461 | 472 | 504 |
| 55 | 765 | 890 | 770 | 660 | 690 | 575 | 490 | 485 | 475 | 509 |
| 56 | 690 | 785 | 645 | 510 | 560 | 535 | 405 | 474 | 455 | 458 |
| 57 | 790 | 770 | 665 | 600 | 545 | 555 | 390 | 389 | 461 | 439 |
| 58 | 825 | 960 | 750 | 660 | 680 | 620 | 510 | 501 | 490 | 534 |
| 59 | 805 | 860 | 635 | 540 | 650 | 555 | 435 | 479 | 457 | 451 |
| 60 | 720 | 705 | 615 | 540 | 625 | 595 | 380 | 379 | 483 | 502 |
| 61 | 735 | 805 | 665 | 535 | 605 | 580 | 430 | 427 | 441 | 498 |
| 62 | 855 | 905 | 700 | 615 | 650 | 615 | 495 | 531 | 448 | 453 |
| 63 | 765 | 945 | 820 | 695 | 750 | 685 | 530 | 511 | 580 | 510 |
| 64 | 750 | 825 | 715 | 580 | 660 | 655 | 505 | 414 | 451 | 542 |
| 65 | 790 | 790 | 695 | 595 | 625 | 510 | 525 | 425 | 471 | 478 |
| 66 | 845 | 995 | 820 | 655 | 675 | 640 | 545 | 474 | 513 | 599 |
| 67 | 870 | 890 | 740 | 615 | 655 | 555 | 490 | 433 | 558 | 590 |
| 68 | 825 | 920 | 640 | 600 | 690 | 575 | 535 | 451 | 484 | 510 |
| 69 | 865 | 935 | 690 | 645 | 640 | 530 | 520 | 476 | 507 | 525 |
| 70 | 860 | 960 | 725 | 615 | 710 | 660 | 525 | 438 | 539 | 646 |
| 71 | 910 | 975 | 775 | 680 | 700 | 770 | 635 | 506 | 526 | 514 |
| 72 | 745 | 815 | 700 | 605 | 615 | 605 | 550 | 455 | 504 | 503 |
| 73 | 810 | 730 | 635 | 535 | 650 | 735 | 535 | 458 | 554 | 535 |
| 74 | 745 | 840 | 730 | 645 | 650 | 775 | 590 | 557 | 561 | 565 |
| 75 | 730 | 775 | 680 | 610 | 610 | 680 | 515 | 537 | 535 | 553 |
| 76 | 745 | 660 | 565 | 520 | 635 | 610 | 450 | 476 | 550 | 562 |
| 77 | 675 | 690 | 635 | 525 | 605 | 635 | 515 | 494 | 516 | 558 |
| 78 | 700 | 725 | 645 | 645 | 580 | 615 | 640 | 516 | 618 | 615 |
| 79 | 765 | 725 | 615 | 640 | 705 | 710 | 590 | 619 | 627 | 620 |
| 80 | 785 | 655 | 600 | 570 | 615 | 670 | 575 | 506 | 602 | 530 |
| 81 | 550 | 590 | 590 | 605 | 505 | 560 | 595 | 539 | 629 | 571 |
| 82 | 790 | 675 | 600 | 625 | 685 | 725 | 695 | 686 | 608 | 568 |
| 83 | 670 | 630 | 640 | 645 | 650 | 695 | 610 | 627 | 579 | 537 |
| 84 | 730 | 615 | 650 | 640 | 645 | 655 | 620 | 611 | 642 | 685 |
| 85 | 700 | 675 | 720 | 695 | 680 | 720 | 615 | 595 | 556 | 648 |
| 86 | 735 | 645 | 620 | 705 | 695 | 650 | 580 | 623 | 541 | 580 |
| 87 | 820 | 685 | 665 | 715 | 715 | 790 | 660 | 658 | 689 | 725 |
| 88 | 670 | 590 | 580 | 550 | 665 | 640 | 605 | 553 | 580 | 605 |
| 89 | 685 | 505 | 525 | 595 | 590 | 560 | 575 | 592 | 559 | 584 |
| 90 | 760 | 625 | 545 | 635 | 635 | 670 | 625 | 672 | 656 | 577 |
| 91 | 740 | 575 | 565 | 595 | 610 | 650 | 595 | 515 | 641 | 533 |
| 92 | 650 | 570 | 575 | 450 | 445 | 670 | 645 | 587 | 634 | 534 |
| 93 | 705 | 550 | 515 | 530 | 485 | 610 | 570 | 565 | 652 | 518 |
| 94 | 690 | 615 | 610 | 560 | 625 | 745 | 540 | 584 | 616 | 505 |
| 95 | 820 | 720 | 650 | 590 | 615 | 770 | 645 | 615 | 755 | 625 |
| 96 | 780 | 640 | 660 | 605 | 540 | 695 | 580 | 615 | 679 | 540 |
| 97 | 800 | 715 | 710 | 605 | 575 | 490 | 485 | 564 | 665 | 665 |
| 98 | 830 | 795 | 765 | 610 | 630 | 775 | 630 | 614 | 760 | 730 |
| 99 | 685 | 680 | 735 | 560 | 580 | 670 | 610 | 556 | 539 | 488 |

## Appendix 5 A note on estimating the population variance $\sigma^2$

### Definition

If $\hat{\theta}$ is a sample statistic used to estimate the parameter $\theta$ and $E(\hat{\theta}) = \theta$ then $\hat{\theta}$ is said to be an **unbiased estimator** of $\theta$.

*Example* If $X_1, X_2, \ldots, X_n$ are independent and identically distributed random variables with

$$E(X_i) = \mu \text{ and } V(X_i) = \sigma^2 \text{ for } i = 1, 2, \ldots, n$$

show that the sample variance $\frac{1}{n-1}\Sigma(X_i - \overline{X})^2$ is an unbiased estimator of $\sigma^2$.

*Solution*

$$\begin{aligned}\Sigma(X_i - \overline{X})^2 &= \Sigma(X_i^2 - 2X_i\overline{X} + \overline{X}^2)\\ &= \Sigma X_i^2 - 2\overline{X}\Sigma X_i + n\overline{X}^2\\ &= \Sigma X_i^2 - 2\overline{X}(n\overline{X}) + n\overline{X}^2\\ &= \Sigma X_i^2 - n\overline{X}^2\end{aligned}$$

Therefore

$$E(\Sigma(X_i - \overline{X})^2) = \Sigma E(X_i^2) - nE(\overline{X}^2)$$

But

$$V(X_i) = E(X_i^2) - (E(X_i))^2 \Rightarrow E(X_i^2) = (E(X_i))^2 + V(X_i) = \mu^2 + \sigma^2$$

Similarly

$$E(\overline{X}^2) = (E(\overline{X}))^2 + V(\overline{X}) = \mu^2 + \frac{\sigma^2}{n}$$

Therefore

$$\begin{aligned}E(\Sigma(X_i - \overline{X})^2) &= \Sigma(\mu^2 + \sigma^2) - n\left(\mu^2 + \frac{\sigma^2}{n}\right)\\ &= n\mu^2 + n\sigma^2 - n\mu^2 - \sigma^2\\ &= (n-1)\sigma^2\end{aligned}$$

and $$E\left(\frac{1}{n-1}\Sigma(X_i - \overline{X})^2\right) = \frac{1}{n-1} \times (n-1)\sigma^2 = \sigma^2$$

Therefore, the sample variance $\frac{1}{n-1}\Sigma(X_i - \overline{X})^2$ is an unbiased estimator of $\sigma^2$.

# Answers

## CHAPTER 1

**Exercise 1.1 (page 3)**

**1** **a** $S = \{0, 1, \ldots, 20\}$
**b** $S = \{\text{perfect, second, imperfect}\}$
**c** $S = \{2, 4, 6, \ldots, 30\}$
**d** $S = \{1, 2, \ldots\}$
**e** $S = \{x: x > 0\}$
**f** $S = \{0, 1, 2, 3, 4, 5\}$
**g** $S = \{x: x \in R\}$ since a person's weight could go up or down

**2** **a** $S = \{0, 1, \ldots, 7000\}$
**b** $S = \{0, 1, \ldots, 1000\}$
**c** $S = \{(x_1, \ldots, x_7): x_1 = 0, 1, \ldots, 1000; \ldots; x_7 = 0, 1, \ldots, 1000\}$

**3** Number of points in the sample space $= 6 \times 6 \times 6 \times 6 \times 6 = 6^5 = 7776$

**Exercise 1.2 (page 6)**

**1** **a** $S = \{1, 2, 3, 4, 5, 6\}$
**b** **(i)** $A = \{1, 3, 5\}$
**(ii)** $B = \{2, 3, 5\}$
**(iii)** $C = \{6\}$
**(iv)** $D = \{5, 6\}$
**(v)** $A$ *or* $B = \{1, 2, 3, 5\}$
**(vi)** $A$ *and* $B = \{3, 5\}$

**2** **a** $S = \{(x, y): x = 1, \ldots, 6; y = 1, \ldots, 6\}$
**b** **(i)** $E = \{(1, 1), (2, 2), (3, 3), (4, 4), (5, 5), (6, 6)\}$
**(ii)** $F = \{(1, 4), (1, 5), (1, 6), (2, 5), (2, 6), (3, 6)\}$
**(iii)** $G = \{(2, 1), (4, 2), (6, 3)\}$
**(iv)** $H = \{(1, 1), (1, 2), (1, 3), (1, 4), (2, 1), (2, 2), (2, 3), (3, 1), (3, 2), (4, 1)\}$
**c** The disjoint pairs of events are $E$ and $F$, $E$ and $G$, $F$ and $G$.

**3** **a** $S$ = {SSSS, SSSU, SSUS, SSUU, SUSS, SUSU, SUUS, SUUU, USSS, USSU, USUS, USUU, UUSS, UUSU, UUUS, UUUU}
**b** **(i)** $A$ = {SSSU, SSUS, SUSS, USSS}
**(ii)** $B$ = {SSSS, SSSU, SSUS, SUSS, USSS}
**(iii)** $C$ = {SSUU, SUSU, SUUS, SUUU, USSU, USUS, USUU, UUSS, UUSU, UUUS, UUUU}
**c** $C = B'$ and $B = C'$

**4** **a** $S$ = {GGG, GGF, GGS, GFG, GFF, GFS, GSG, GSF, GSS, FGG, FGF, FGS, FFG, FFF, FFS, FSG, FSF, FSS, SGG, SGF, SGS, SFG, SFF, SFS, SSG, SSF, SSS}
**b** **(i)** {GGG, GGS, GSG, GSS, SGG, SGS, SSG, SSS}
**(ii)** {SSS}
**(iii)** {GSS, SGS, SSG}
**(iv)** {GGG, FFF, SSS}

**5** **a** $S = \{x: x \geq 0\}$
**b** **(i)** $\{x: x \geq 5\}$
**(ii)** $\{0\}$ (which is not the empty set)
**(iii)** $\{x: 0 \leq x < 10\}$
**(iv)** $\{x: 2 \leq x \leq 8\}$
**c** **(i)** $\{x: 0 \leq x < 5\}$
**(ii)** $\{x: x > 0\}$
**(iii)** $\{x: x \geq 10\}$
**(iv)** $\{x: 0 \leq x < 2 \text{ or } x > 8\}$

**Exercise 1.3 (page 9)**

**1** **a** $P(A) = \frac{3}{6} = \frac{1}{2}$ $\quad P(B) = \frac{3}{6} = \frac{1}{2}$
$P(C) = \frac{1}{6}$ $\quad P(D) = \frac{2}{6} = \frac{1}{3}$
**b** $P(A \text{ or } B) = \frac{4}{6} = \frac{2}{3}$ $\quad P(A \text{ and } B) = \frac{2}{6} = \frac{1}{3}$
$P(A \text{ or } B) = P(A) + P(B) - P(A \text{ and } B)$
$= \frac{1}{2} + \frac{1}{2} - \frac{1}{3} = \frac{2}{3}$

**2** $P(E) = \frac{6}{36} = \frac{1}{6}$ $\quad P(F) = \frac{6}{36} = \frac{1}{6}$
$P(G) = \frac{3}{36} = \frac{1}{12}$ $\quad P(H) = \frac{10}{36} = \frac{5}{18}$

**3** Let $X$ = 'household gets paper X' and $Y$ = 'household gets paper Y'.
$P(X) = 0.70$, $P(Y) = 0.80$, $P(X \text{ and } Y) = 0.60$
**a** $P(X \text{ or } Y) = P(X) + P(Y) - P(X \text{ and } Y)$
$= 0.90$
90% of households get at least one paper
**b** P(household gets neither paper)
= 1 − P(household gets at least one paper)
= 0.10
10% of households get neither paper

**4** Let $E$ = 'Elsie is at home' and $F$ = 'Frank is at home'.
$P(E) = 0.50$, $P(F) = 0.65$
$P(E \text{ or } F) = 1 - \text{P(neither is at home)}$
$= 1 - 0.10 = 0.90$
Now $P(E \text{ or } F) = P(E) + P(F) - P(E \text{ and } F)$
i.e. $0.90 = 0.50 + 0.65 - P(E \text{ and } F)$
i.e. $P(E \text{ and } F) = 0.25$
Both Elsie and Frank are at home 25% of the times that I call.

**5** $P(E \text{ or } F) = P(E) + P(F) - P(E \text{ and } F)$
When $E$ and $F$ are disjoint, $P(E \text{ and } F) = 0$, so the general addition rule becomes
$P(E \text{ and } F) = P(E) + P(F)$

### Exercise 1.4 (page 12)

**1 a** No. of outcomes = $2 \times 2 \times 2 \times 2 = 16$

$S$ = {HHHH, HHHT, HHTH, HHTT, HTHH, HTHT, HTTH, HTTT, THHH, THHT, THTH, THTT, TTHH, TTHT, TTTH, TTTT}

All these outcomes have probability $\frac{1}{16}$.

**b (i)** $\frac{15}{16}$ **(ii)** $\frac{6}{16} = \frac{3}{8}$ **(iii)** $\frac{5}{16}$

**2 a** $\frac{1}{5}$

**b** No. of outcomes = $5 \times 5 = 25$

**(i)** No. of favourable outcomes = 1, so probability = $\frac{1}{25}$

**(ii)** No. of favourable outcomes = $4 \times 4$, so probability = $\frac{16}{25}$

**(iii)** No. of favourable outcomes = $25 - (1 + 16) = 8$, so probability = $\frac{8}{25}$

**3 a (i)** $\frac{18}{38} = \frac{9}{19}$ **(ii)** $\frac{18}{38} = \frac{9}{19}$

**(iii)** $\frac{2}{38} = \frac{1}{19}$ **(iv)** $\frac{7}{38}$

**b (i)** No. of outcomes = $38 \times 38 = 38^2$

**(ii)** No. of favourable outcomes = $2 \times 2 = 4$, so probability = $\frac{4}{38^2} = 0.002\,77$

**4 a (i)** $P(E) = 0.5$ **(ii)** $P(F) = 0.4$

**(iii)** $P(E \text{ and } F) = P(0, 6) = 0.2$

**(iv)** $P(E \text{ or } F) = P(E) + P(F) - P(E \text{ and } F) = 0.7$

**b** $Y$ can be any integer between 0 and 99. All these outcomes are equally likely.

**(i)** $G = \{0, 2, 4, 6, \ldots, 98\}$, $P(G) = 0.5$

**(ii)** $H = \{0, 5, 10, \ldots, 95\}$, $P(H) = 0.2$

**(iii)** $M' = \{0, 10, 20, \ldots, 90\}$, $P(M') = 0.1$, so $P(M) = 1 - P(M') = 0.9$

**(iv)** $G \text{ and } H = \{0, 10, 20, \ldots, 90\}$, so $P(G \text{ and } H) = 0.1$

**(v)** $G \text{ and } M = \{2, 4, 6, 8, 12, \ldots, 98\}$, so $P(G \text{ and } M) = 0.4$

**(vi)** $H \text{ and } M = \{5, 15, \ldots, 95\}$ so $P(H \text{ and } M) = 0.1$

**c** $G \text{ and } H = M'$

### Exercise 1.5A (page 15)

**1 a** $5! = 5 \times 4 \times 3 \times 2 \times 1 = 120$

**b** $6! = 6 \times 5! = 720$

**c** $10! = 10 \times 9 \times 8 \times 7 \times 6! = 3\,628\,800$

**d** $20! = 2\,432\,902\,008\,176\,640\,000$

**2 a** ${}^4P_3 = 4 \times 3 \times 2 = 24$

**b** ${}^{10}P_5 = 10 \times 9 \times 8 \times 7 \times 6 = 30\,240$

**c** ${}^7P_2 = 7 \times 6 = 42$

**d** ${}^{12}P_4 = 12 \times 11 \times 10 \times 9 = 11\,880$

**e** ${}^8P_6 = 8 \times 7 \times 6 \times 5 \times 4 \times 3 = 20\,160$

**3 a** $\binom{4}{3} = \frac{4 \times 3 \times 2}{1 \times 2 \times 3} = 4$

**b** $\binom{10}{5} = \frac{10 \times 9 \times 8 \times 7 \times 6}{1 \times 2 \times 3 \times 4 \times 5} = 252$

**c** $\binom{7}{2} = \frac{7 \times 6}{1 \times 2} = 21$

**d** $\binom{12}{4} = \frac{12 \times 11 \times 10 \times 9}{1 \times 2 \times 3 \times 4} = 495$

**e** $\binom{8}{6} = \frac{8 \times 7 \times 6 \times 5 \times 4 \times 3}{1 \times 2 \times 3 \times 4 \times 5 \times 6} = 28$

**4** $5! = 5 \times 4 \times 3 \times 2 \times 1 = 120$ different ways

**5** $3 \times 2 \times 4 = 24$ different routes

**6 a** $3! = 6$ ways

XYZ, XZY, YXZ, YZX, ZXY, ZYX

**b (i)** $\frac{1}{3}$ **(ii)** $\frac{1}{3}$ **(iii)** $\frac{1}{3}$

**7 a** $\binom{1000}{5}$

**b** There are $\binom{997}{5}$ different ways in which all the prizes are won by the 997 tickets bought by other people. So

$$P(\text{win no prize}) = \frac{\binom{997}{5}}{\binom{1000}{5}}$$

$$= \frac{997 \times 996 \times 995 \times 994 \times 993}{1000 \times 999 \times 998 \times 997 \times 996}$$

$= 0.985\,06$

P(win at least one prize) = $0.014\,94$

### Exercise 1.5B (page 16)

**1 a** ${}^5P_1 = 5$

${}^5P_2 = 5 \times 4 = 20$

${}^5P_3 = 5 \times 4 \times 3 = 60$

${}^5P_4 = 5 \times 4 \times 3 \times 2 = 120$

${}^5P_5 = 5 \times 4 \times 3 \times 2 \times 1 = 120$

**b** ${}^nP_n = n!$

**c** ${}^nP_{n-1} = {}^nP_n$

**2 a (i)** $\binom{3}{r} = 1, 3, 3, 1$

**(ii)** $\binom{4}{r} = 1, 4, 6, 4, 1$

**(iii)** $\binom{5}{r} = 1, 5, 10, 10, 5, 1$

**b (i)** $\binom{n}{0} = 1$ **(ii)** $\binom{n}{1} = n$

**(iii)** $\binom{n}{n-1} = n$ **(iv)** $\binom{n}{n} = 1$

c $\binom{n}{n-r} = \dfrac{n!}{(n-r)![n-(n-r)]!}$

$= \dfrac{n!}{(n-r)!r!}$

$= \binom{n}{r}$

d $\binom{n}{0} + \binom{n}{1} + \cdots + \binom{n}{n} = 2^n$

3 a No. of different ways of answering $= 5 \times 5 \times \ldots \times 5 = 5^{10} = 9\,765\,625$

b No. of ways of answering all wrongly $= 4 \times 4 \times \ldots \times 4 = 4^{10} = 1\,048\,576$

P(answer all wrongly when guessing)

$= \dfrac{1\,048\,576}{9\,765\,625} = \left(\left(\dfrac{4}{5}\right)^{10}\right) = 0.1074$

4 a $6 \times 6 \times 6 \times 6 \times 6 = 6^5 = 7776$ outcomes

b P(same score on all dice)

$= \dfrac{6}{7776} = 0.000\,771\,6$

c There are $\binom{5}{4} = 5$ different ways of choosing four dice on which to obtain 1s. For each of these choices, there are five different choices of score on the fifth die (2, 3, …, 6). So, there are $5 \times 5 = 25$ different ways to obtain exactly four 1s from five dice. Thus

P(exactly four 1s) $= \dfrac{25}{7776} = 0.003\,215$

d There are also 25 ways of obtaining exactly four 2s, four 3s, …, four 6s. These events are disjoint, and so P(same score on exactly four dice)

$= \dfrac{6 \times 25}{7776} = 0.019\,29\ (01)$

e P(same score on at least four dice)

= P(same score on exactly four dice) + P(same score on all five dice)

= 0.020 06 (17)

5 a $\binom{25}{5} = 53\,130$ b $\binom{20}{5} = 15\,504$

c $5! = 120$ d $\frac{1}{5}$

6 a $\binom{100}{5}$ b $\binom{60}{3}$ c $\binom{40}{2}$

d Denominator is number of possible outcomes. Numerator is number of outcomes in which exactly three girls (and two boys) are chosen.

e $\dfrac{\binom{60}{x}\binom{40}{5-x}}{\binom{100}{5}}$

### Exercise 1.6A (page 21)

1 a $P(F|E) = \frac{7}{19}$ b $P(F|E') = \frac{8}{19}$

c $P(E \text{ and } F) = P(F|E)P(E)$

$= \frac{7}{19} \times \frac{8}{20} = 0.1474$ or $\frac{14}{19}$

2 a $P(F|E) = \frac{11}{21}$ b $P(F|E') = \frac{10}{21}$

c $P(E' \text{ and } F') = P(F'|E')P(E')$

$= \left(1 - \frac{10}{21}\right) \times \frac{10}{20} = 0.2619$

3 a $\frac{169}{221} = \frac{13}{17}$ b $\frac{17}{221} = \frac{1}{13}$ c $\frac{44}{221}$ d $\frac{15}{169}$

e $\frac{26}{169} = \frac{2}{13}$ f $\frac{15}{17}$ g $\frac{26}{44} = \frac{13}{22}$

4 a $P(H) = \frac{1}{2}$ b $P(TH) = P(T) \times P(H) = \frac{1}{4}$

5 a P(both engines fail) $= 0.01 \times 0.01 = 0.0001$

b P(neither engine fails) $= 0.99 \times 0.99 = 0.9801$

c P(exactly one engine fails) $= 1 - (0.0001 + 0.9801) = 0.0198$

6 P(M *and* SH approve)

= P(M approves | SH approves) × P(SH approves)

$= 0.9 \times 0.8 = 0.72$

7 a $P(E \text{ or } F) = P(E) + P(F) = 0.9$

b $P(E \text{ or } F) = P(E) + P(F) - P(E \text{ and } F)$

$= P(E) + P(F) - P(E) \times P(F)$

$= 0.72$

c $P(E \text{ or } F) = P(E) + P(F) - P(E \text{ and } F)$

$= P(E) + P(F) - P(E|F) \times P(F)$

$= 0.66$

8 a P(A wins both games) $= \frac{3}{4} \times \frac{2}{5} = 0.3$

b P(B wins both games) $= \frac{1}{4} \times \frac{1}{3} = 0.0833$

9 Let $R$ = 'offspring has red flowers' = 'offspring has genotype *cc*'

parents are *CC*, *CC* : $P(R) = 0 \times 0 = 0$

parents are *CC*, *Cc* : $P(R) = 0 \times \frac{1}{2} = 0$

parents are *CC*, *cc* : $P(R) = 0 \times 1 = 0$

parents are *Cc*, *Cc* : $P(R) = \frac{1}{2} \times \frac{1}{2} = \frac{1}{4}$

parents are *Cc*, *cc* : $P(R) = \frac{1}{2} \times 1 = \frac{1}{2}$

parents are *cc*, *cc* : $P(R) = 1 \times 1 = 1$

### Exercise 1.6B (page 22)

1 a $\frac{1}{6}$ b 0 c $\frac{1}{6}$

d $\frac{1}{6} \times \frac{1}{6} = \frac{1}{36}$ e $\frac{1}{36}$

2 $E$, $F$ disjoint means $P(E \text{ and } F) = 0$.

$P(E) > 0$, $P(F) > 0$ means $P(E) \times P(F) > 0$.

Hence, $P(E \text{ and } F) \neq P(E) \times P(F)$.

3 a Assuming independence,

(i) P(X *and* Y attend) $= 0.8 \times 0.6 = 0.48$

(ii) P(neither X nor Y attends) $= 0.2 \times 0.4 = 0.08$

**b** P(X attends, Y does not)
$= 0.8 \times 0.4 = 0.32$
P(X does not attend, Y attends)
$= 0.2 \times 0.6 = 0.12$
P(exactly one of X, Y attends)
$= 0.32 + 0.12 = 0.44$ (disjoint events)

**4** Let A denote a teenager who attends, D a teenager who does not attend.

**a** $S = \{$AAAA, AAAD, AADA, AADD, ADAA, ADAD, ADDA, ADDD, DAAA, DAAD, DADA, DADD, DDAA, DDAD, DDDA, DDDD$\}$

**b** The corresponding probabilities are
0.6561, 0.0729, 0.0729, 0.0081,
0.0729, 0.0081, 0.0081, 0.0009,
0.0729, 0.0081, 0.0081, 0.0009,
0.0081, 0.0009, 0.0009, 0.0001

**c** $P(X = 0) = 0.6561$, $P(X = 1) = 0.2916$
$P(X = 2) = 0.0486$, $P(X = 3) = 0.0036$
$P(X = 4) = 0.0001$

### Exercise 1.7B (page 26)

**1 a** Probabilities are:

| | | | |
|---|---|---|---|
| Stage 1 | boy | 0.52 | |
| | girl | 0.48 | |
| Stage 2 | (boy) | colour blind | 0.05 |
| | | not colour blind | 0.95 |
| | (girl) | colour blind | 0.0025 |
| | | not colour blind | 0.9975 |

**b** P(child is colour blind)
$= (0.05 \times 0.52) + (0.0025 \times 0.48) = 0.0272$
i.e. 2.72% of children are colour blind

**c** P(girl | colour blind)
$= \dfrac{0.0025 \times 0.48}{(0.05 \times 0.52) + (0.0025 \times 0.48)}$
$= 0.0441$
i.e. 4.41% of colour-blind children are girls

**2 a** P(defective)
$= (0.01 \times 0.4) + (0.01 \times 0.3) + (0.04 \times 0.3)$
$= 0.019$
i.e. 1.9% of items are defective

**b** P(C | defective)
$= \dfrac{0.04 \times 0.3}{(0.01 \times 0.4) + (0.01 \times 0.3) + (0.04 \times 0.3)}$
$= 0.6316$
i.e. 63.16% of defective items are from line C

**3 a** P(called to home)
$= (0.2 \times 0.1) + (0.3 \times 0.65) + (0.5 \times 0.25)$
$= 0.34$
i.e. he is called to homes of 34% of clients in the course of a year

**b** Excellent $\frac{20}{340} = \frac{1}{17}$, adequate $\frac{195}{340} = \frac{39}{68}$,
inadequate $\frac{125}{340} = \frac{25}{68}$

**4 a** P(final is decided in two sets)
$= (0.7 \times 0.4) + (0.9 \times 0.6) = 0.82$

**b** P(final is decided in three sets)
$= 1 -$ P(it is decided in two sets) $= 0.18$

**c** P(A wins | final is decided in two sets)
$= \dfrac{0.7 \times 0.4}{(0.7 \times 0.4) + (0.9 \times 0.6)} = 0.3415$

**5 a** P(first) $= 0.6 \times 0.6 = 0.36$
P(second)
$= (0.6 \times 0.3) + (0.3 \times 0.6) + (0.3 \times 0.3)$
$= 0.45$
P(reject) $= 1 - (0.36 + 0.45) = 0.19$

**b** $\dfrac{0.1 \times 0.1}{0.19} = 0.0526$

**c** The inspectors would be likely to have very similar opinions, i.e. they do not rate ornaments independently.

### Chapter 1 Review Exercise (page 31)

**1 a** $6 \times 6 \times 6 = 216$ possible outcomes

**b** No. of favourable outcomes = 6, so
P(same score on all three dice) $= \frac{1}{36}$

**c** No. of favourable outcomes
$= {}^6P_3 = 6 \times 5 \times 4 = 120$
P(diff. scores on all three dice) $= \frac{120}{216}$

**d** No. of favourable outcomes
$= 216 - (6 + 120) = 90$
P(two scores the same) $= \frac{90}{216}$

**2 a** P(C is chosen) $= \frac{3}{20} = 0.15$

**b** There are $\binom{20}{3}$ ways of choosing the three pupils. In $\binom{18}{3}$ of these combinations, neither C nor M is chosen. Hence
P(neither C nor M chosen) $= 0.7158$
P(at least one of them is chosen)
$= 1 - 0.7158 = 0.2842$

**c** P(C *or* M) = P(C) + P(M) − P(C *and* M)
$0.2842 = 0.15 + 0.15 -$ P(C *and* M)
P(C *and* M) $= 0.0158$

**3 a (i)** P(long pollen) $= \frac{5224}{6952} = 0.7514$

**(ii)** P(long pollen | purple flowers)
$= \frac{4831}{5221} = 0.9253$

**(iii)** P(long pollen | red flowers)
$= \frac{393}{1731} = 0.2270$

**b** If these two characteristics were independent, then (in the population) all three of the above probabilities should be equal. This seems unlikely on the basis of this sample.

**4** P(student passes course)
$= 0.90 \times 0.88 = 0.792$
P(student fails course)
$= 1 - 0.792 = 0.208$
i.e. 20.8% of students fail this course.

**5** **a** P(no sale)
= P(no sale | 1 client) × P(1 client)
+ P(no sale | 2 clients) × P(2 clients)
$= \left(\frac{2}{3} \times \frac{1}{4}\right) + \left(\frac{4}{9} \times \frac{3}{4}\right) = \frac{1}{2}$

**b** P(at least one sale) $= 1 - \frac{1}{2} = \frac{1}{2}$

**6** **a** P(sale within one month)
$= (0.95 \times 0.2) + (0.8 \times 0.7) + (0.5 \times 0.1)$
$= 0.8$

**b** P(unpopular | sale within one month)
$$= \frac{0.5 \times 0.1}{(0.95 \times 0.2) + (0.8 \times 0.7) + (0.5 \times 0.1)}$$
$= \frac{1}{16}$

# CHAPTER 2

### Exercise 2.1A (page 34)

**1** **a**

| $t$ | 2 | 3 | 4 | 5 | 6 | 7 | 8 | 9 | 10 | 11 | 12 |
|---|---|---|---|---|---|---|---|---|---|---|---|
| p($t$) | $\frac{1}{36}$ | $\frac{2}{36}$ | $\frac{3}{36}$ | $\frac{4}{36}$ | $\frac{5}{36}$ | $\frac{6}{36}$ | $\frac{5}{36}$ | $\frac{4}{36}$ | $\frac{3}{36}$ | $\frac{2}{36}$ | $\frac{1}{36}$ |

**b** $E(T) = 7$, $V(T) = \frac{35}{6}$

**2** **a**

| $y$ | 1 | 2 | 3 | 4 | 5 |
|---|---|---|---|---|---|
| p($y$) | $\frac{1}{5}$ | $\frac{1}{5}$ | $\frac{1}{5}$ | $\frac{1}{5}$ | $\frac{1}{5}$ |

**b** $E(Y) = 3$, $V(Y) = 2$

**c**

| $y$ | 1 | 2 | 3 | 4 | 5 |
|---|---|---|---|---|---|
| $P(Y \le y)$ | $\frac{1}{5}$ | $\frac{2}{5}$ | $\frac{3}{5}$ | $\frac{4}{5}$ | 1 |

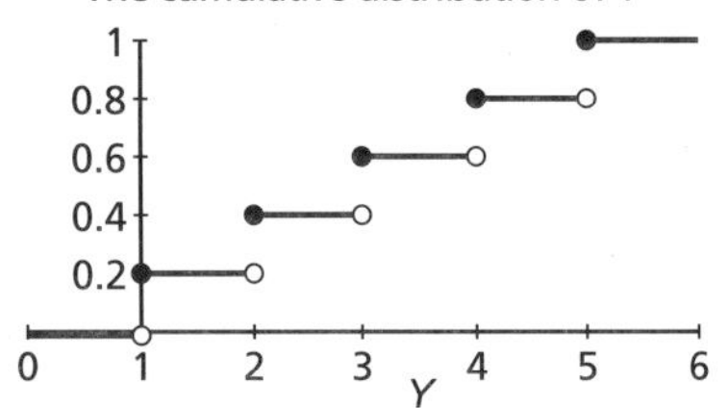


**3** **a**

| $x$ | 0 | 1 | 2 | 3 |
|---|---|---|---|---|
| p($x$) | $\frac{2}{42}$ | $\frac{15}{42}$ | $\frac{20}{42}$ | $\frac{5}{42}$ |

**b** $E(X) = \frac{5}{3}$, $V(X) = \frac{5}{9}$

**4**

| $z$ | 1 | 2 | 3 | 4 | 5 | 7 | 8 | 9 | 10 | 11 | 12 |
|---|---|---|---|---|---|---|---|---|---|---|---|
| p($z$) | $\frac{1}{6}$ | $\frac{1}{6}$ | $\frac{1}{6}$ | $\frac{1}{6}$ | $\frac{1}{6}$ | $\frac{1}{36}$ | $\frac{1}{36}$ | $\frac{1}{36}$ | $\frac{1}{36}$ | $\frac{1}{36}$ | $\frac{1}{36}$ |

$E(Z) = \frac{49}{12}$, $V(Z) = \frac{385}{48}$

**5** **a**

| $x$ | 1 | 2 | 3 | 4 |
|---|---|---|---|---|
| p($x$) | 0.1 | 0.2 | 0.3 | 0.4 |

**b** $E(X) = 3$, $V(X) = 1$, $sd(X) = 1$

**6** **a** $k = \frac{1}{32}$

**b** $E(X) = \frac{5}{2}$, $V(X) = \frac{5}{4}$

**c**

| $x$ | 0 | 1 | 2 | 3 | 4 | 5 |
|---|---|---|---|---|---|---|
| $P(X \le x)$ | $\frac{1}{32}$ | $\frac{6}{32}$ | $\frac{16}{32}$ | $\frac{26}{32}$ | $\frac{31}{32}$ | 1 |

### Exercise 2.1B (page 36)

**1** **a** Tree diagram with outcome probabilities:

| Outcome | Probability | $z$ |
|---|---|---|
| FFF | 0.8×0.6×0.4 = 0.192 | 0 |
| FFS | 0.8×0.6×0.6 = 0.288 | 1 |
| FSF | 0.8×0.4×0.4 = 0.128 | 1 |
| FSS | 0.8×0.4×0.6 = 0.192 | 2 |
| SFF | 0.2×0.6×0.4 = 0.048 | 1 |
| SFS | 0.2×0.6×0.6 = 0.072 | 2 |
| SSF | 0.2×0.4×0.4 = 0.032 | 2 |
| SSS | 0.2×0.4×0.6 = 0.048 | 3 |

**b**

| $z$ | 0 | 1 | 2 | 3 |
|---|---|---|---|---|
| p($z$) | 0.192 | 0.464 | 0.296 | 0.048 |

**c** $E(Z) = 1.2$, $V(Z) = 0.64$, $sd(Z) = 0.8$

**2** **a**

| $x$ | 1 | 2 | 3 | 4 | 5 |
|---|---|---|---|---|---|
| p($x$) | $\frac{3}{25}$ | $\frac{4}{25}$ | $\frac{5}{25}$ | $\frac{6}{25}$ | $\frac{7}{25}$ |

**b** $E(X) = \frac{17}{5}$, $V(X) = \frac{46}{25}$

**3** **a**

| $x$ | 0 | 1 | 2 |
|---|---|---|---|
| $P(X = x)$ | $(1 - p)^2$ | $2p(1 - p)$ | $p^2$ |

**c** $p = \frac{1}{3}$, $E(X) = \frac{2}{3}$, $V(X) = \frac{4}{9}$

**4 a** $V(Y) = E(Y^2) - \mu^2 = 4k + 0 + 4k - 0^2 = 8k$

**b (i)** When $k = \frac{1}{8}$, $V(Y) = 1$

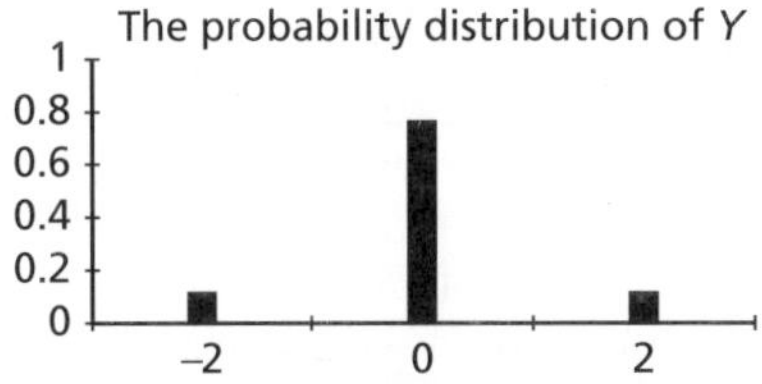

The probability distribution of Y
1
0.8
0.6
0.4
0.2
0
–2
0
2


**(ii)** When $k = \frac{1}{3}$, $V(Y) = 2\frac{2}{3}$

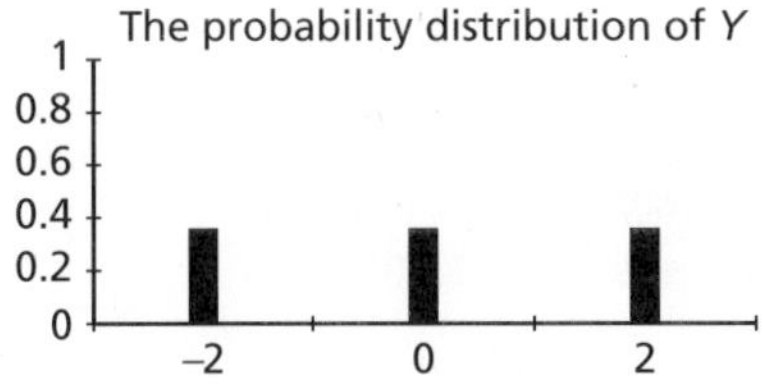

The probability distribution of Y
1
0.8
0.6
0.4
0.2
0
–2
0
2


**(iii)** When $k = \frac{1}{2}$, $V(Y) = 4$

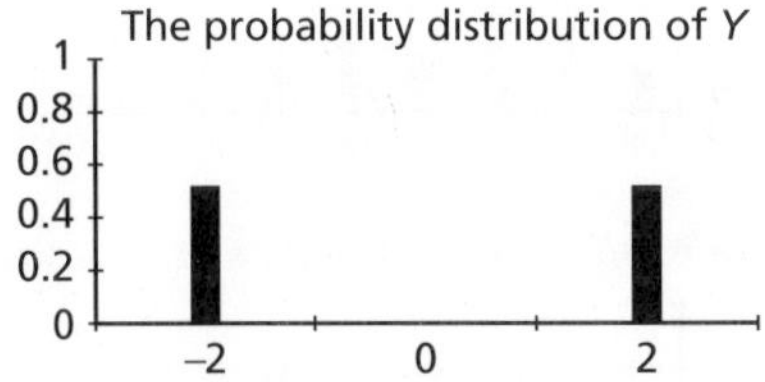

The probability distribution of Y
1
0.8
0.6
0.4
0.2
0
–2
0
2


**5 a** $E(X) = \sum_{i=1}^{n} x_i p(x_i) = \frac{1}{n}\sum_{i=1}^{n} x_i$

$= \frac{1}{n} \times \frac{1}{2}n(n+1) = \frac{1}{2}(n+1)$

$E(X^2) = \sum_{i=1}^{n} x_i^2 p(x_i) = \frac{1}{n}\sum_{i=1}^{n} x_i^2$

$= \frac{1}{6}(n+1)(2n+1)$

$V(X) = \frac{1}{6}(n+1)(2n+1) - \frac{1}{4}(n+1)^2$

$= \frac{1}{12}(n^2 - 1)$

**b** $n = 6$, $E(X) = \frac{7}{2}$, $V(X) = \frac{35}{12}$

**c** $E(X) = \frac{1}{2}$, $V(X) = \frac{35}{12}$

**d** $E(X) = k + \frac{1}{2}(n+1)$

$V(X) = \frac{1}{12}(n^2 - 1)$

**e** $E(X) = \frac{9}{2}$, $V(X) = \frac{33}{4}$

**Exercise 2.2 (page 41)**

**1 a (i)**

| $w$ | 3 | 4 | 5 | 6 |
|---|---|---|---|---|
| $p(w)$ | $\frac{1}{4}$ | $\frac{1}{4}$ | $\frac{1}{4}$ | $\frac{1}{4}$ |

**(ii)**

| $w$ | 3 | 6 | 9 | 12 |
|---|---|---|---|---|
| $p(w)$ | $\frac{1}{4}$ | $\frac{1}{4}$ | $\frac{1}{4}$ | $\frac{1}{4}$ |

**(iii)**

| $w$ | 5 | 8 | 11 | 14 |
|---|---|---|---|---|
| $p(w)$ | $\frac{1}{4}$ | $\frac{1}{4}$ | $\frac{1}{4}$ | $\frac{1}{4}$ |

**b, c**

**(i)** $E(W) = \frac{9}{2}$, $V(W) = \frac{5}{4}$

**(ii)** $E(W) = \frac{15}{2}$, $V(W) = \frac{45}{4}$

**(iii)** $E(W) = \frac{19}{2}$, $V(W) = \frac{45}{4}$

**2 a** $E(W) = 2$, $V(W) = 4$

**b** $E(W) = 20$, $V(W) = 64$

**c** $E(W) = 5$, $V(W) = 16$

**3 a (i)**

| $x + y$ | 0 | 1 | 2 |
|---|---|---|---|
| $p(x + y)$ | $\frac{1}{4}$ | $\frac{1}{2}$ | $\frac{1}{4}$ |

**(ii)**

| $x - y$ | –1 | 0 | 1 |
|---|---|---|---|
| $p(x - y)$ | $\frac{1}{4}$ | $\frac{1}{2}$ | $\frac{1}{4}$ |

**(iii)**

| $3x - 2y$ | –2 | 0 | 1 | 3 |
|---|---|---|---|---|
| $p(3x - 2y)$ | $\frac{1}{4}$ | $\frac{1}{4}$ | $\frac{1}{4}$ | $\frac{1}{4}$ |

**b, c**

**(i)** $E(X + Y) = 1$, $V(X + Y) = \frac{1}{2}$

**(ii)** $E(X - Y) = 0$, $V(X - Y) = \frac{1}{2}$

**(iii)** $E(3X - 2Y) = \frac{1}{2}$, $V(3X - 2Y) = \frac{13}{4}$

**4 a**

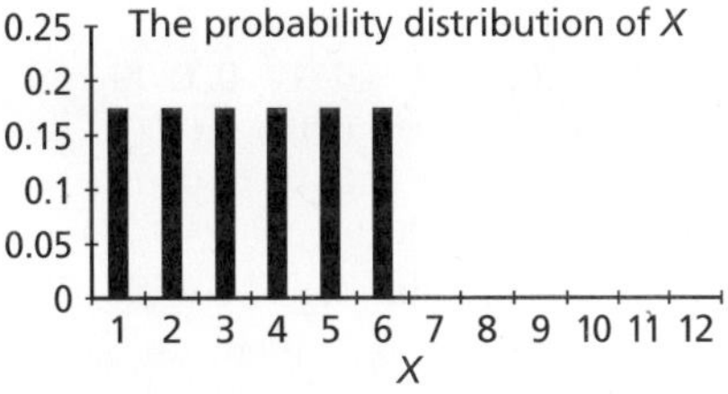

The probability distribution of X
0.25
0.2
0.15
0.1
0.05
0
1 2 3 4 5 6 7 8 9 10 11 12
X


$E(X) = \frac{7}{2}$, $V(X) = \frac{35}{12}$

**b** Same bar diagram, mean and variance as in **a**

**c**

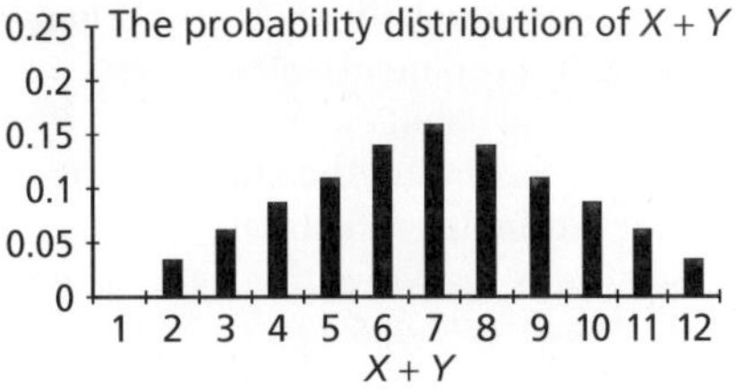

The probability distribution of X + Y
0.25
0.2
0.15
0.1
0.05
0
1 2 3 4 5 6 7 8 9 10 11 12
X + Y


$E(X + Y) = 7$, $V(X + Y) = \frac{35}{6}$

**5** **a** $E(3X - 2) = 58$
$V(3X - 2) = 144$, $sd(3X - 2) = 12$
**b** $E(X - Y) = 5$
$V(X - Y) = 25$, $sd(X - Y) = 5$

**6** **a** $E(X + Y) = 25$, $sd(X + Y) = 5$
**b** $E(X - Y) = 5$, $sd(X - Y) = 5$
**c** $E(4X - 3Y) = 30$
$sd(4X - 3Y) = \sqrt{337} \approx 18.36$
**d** $E(2X + Y - 5) = 35$
$sd(2X + Y - 5) = \sqrt{73} \approx 8.54$

**7** **a**

| $xy$ | 1 | 2 | 3 | 4 | 5 | 6 | 8 | 9 | 10 |
|---|---|---|---|---|---|---|---|---|---|
| $p(xy)$ | $\frac{1}{36}$ | $\frac{2}{36}$ | $\frac{2}{36}$ | $\frac{3}{36}$ | $\frac{2}{36}$ | $\frac{4}{36}$ | $\frac{2}{36}$ | $\frac{1}{36}$ | $\frac{2}{36}$ |
| $xy$ | 12 | 15 | 16 | 18 | 20 | 24 | 25 | 30 | 36 |
| $p(xy)$ | $\frac{4}{36}$ | $\frac{2}{36}$ | $\frac{1}{36}$ | $\frac{2}{36}$ | $\frac{2}{36}$ | $\frac{2}{36}$ | $\frac{1}{36}$ | $\frac{2}{36}$ | $\frac{1}{36}$ |

**b, c** $E(XY) = \frac{49}{4} = E(X)E(Y)$

### Exercise 2.3A (page 44)

**1** **a** 0.2668 **b** 0.0735 **c** 0.0004

**2** **a** 0.0135, 0.0725, 0.1756, 0.2522
**b** 0.0002, 0.0030, 0.0161, 0.0537
**c** 0.0003, 0.0030, 0.0150, 0.0468

**3**

| $x$ | $p(x)$ | $y$ | $p(y)$ |
|---|---|---|---|
| 0 | 0.16807 | 0 | 0.00243 |
| 1 | 0.36015 | 1 | 0.02835 |
| 2 | 0.30870 | 2 | 0.13230 |
| 3 | 0.13230 | 3 | 0.30870 |
| 4 | 0.02835 | 4 | 0.36015 |
| 5 | 0.00243 | 5 | 0.16807 |

The probabilities are reversed.

**4** **a** 0.0007, 0.0078, 0.0413, 0.1239
**b** 0.0001, 0.0010, 0.0068, 0.0277
**c** 0.0000, 0.0000, 0.0000, 0.0000

**5** **a** 0.1934 **b** 0.0916

**6** Binomial is suitable for **a**, **b** with $n = 4$, $p = 0.5$, and **e** assuming there are no twins etc., $n = 25$, $p = \frac{1}{7}$.
**c** No, $n$ is not constant.
**d** No, $p$ changes as cards are dealt.
**f** No, the number of trials $n$ is not fixed.
**g** No, individuals in each group may not be acting independently.

**7** **a** $P(X \leq 2) = 0.9885$
We are assuming the stock pile is so large that $p$ remains effectively constant as the sample is drawn without replacement.
**b** $\mu = 0.5$, $\sigma^2 = 0.475$

**8** **a** 0.1256
We are assuming that the population is so large that sampling without replacement does not effectively change $p$.
**b** $\mu = 8$, $\sigma \approx 2.19$

### Exercise 2.3B (page 45)

**1** **a** 0.0778, 0.2592, 0.3456, 0.2304
**b** 0.0016, 0.0172, 0.0774, 0.1935
**c** 0.0008, 0.0083, 0.0407, 0.1160

**2** **a** 0.0002, 0.0034, 0.0217, 0.0808
**b** 0.0007, 0.0102, 0.0596, 0.1852
**c** 0.0000, 0.0000, 0.0004, 0.0031

**3** **a** 0.2734 **b** 0.0039

**4** **a** 0.1 **b** 0.4095 **c** 0.8784

**5** **a** 0.0584 **b** $\mu = 2.5$, $\sigma^2 = 1.875$

**6** **a** $X \sim \text{Bin}(18, 0.8)$, $\{0, 1, 2, \ldots, 18\}$
**b** $P(X > 16) = 1 - P(X \leq 16) = 0.0991$
or $P(X \leq 1 \mid p = 0.2)$
**c** $\mu = 14.4$, $\sigma \approx 1.697$

**7** **a** The probability distribution of $X$ when $p = 0.25$

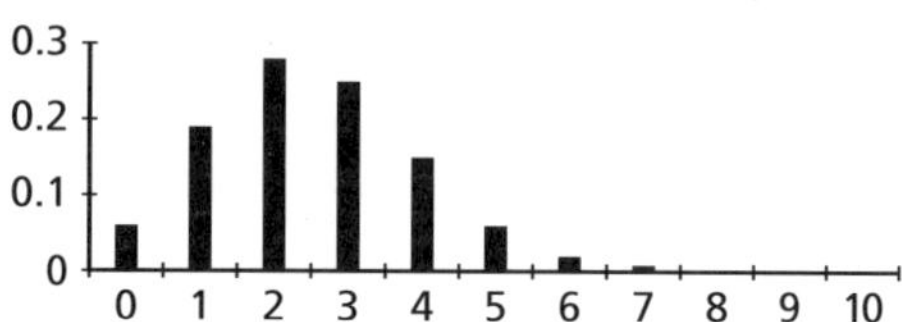


**b** The probability distribution of $X$ when $p = 0.5$

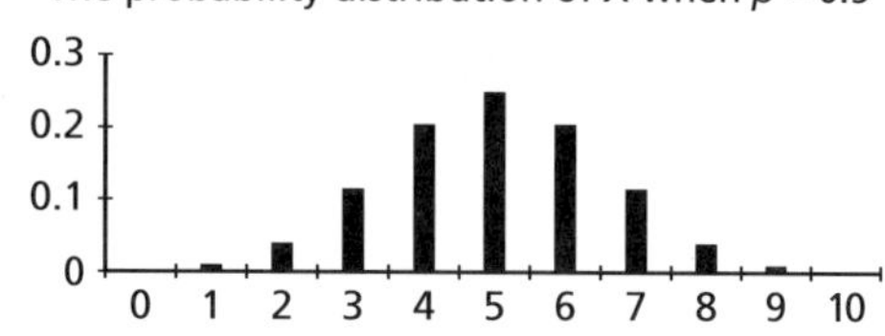


**c** The probability distribution of $X$ when $p = 0.75$

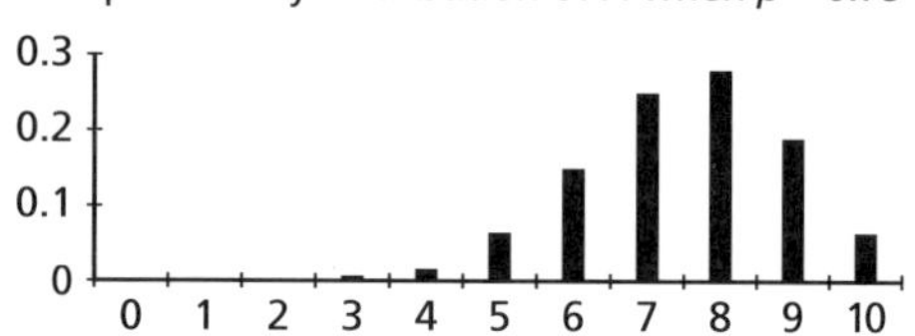


**b** is symmetrical about $X = 5$.
**a** displays positive skew.
**c** displays negative skew.
**c** is the reflection of **a** about $X = 5$.

**8** **a** The probability distribution of $X$ when $n = 4$

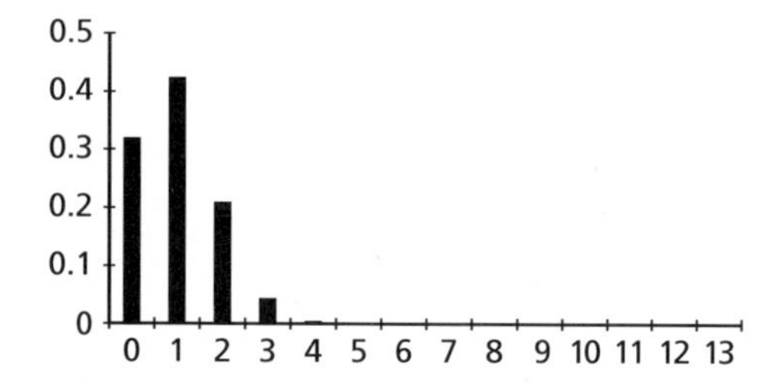

**b** The probability distribution of $X$ when $n = 10$

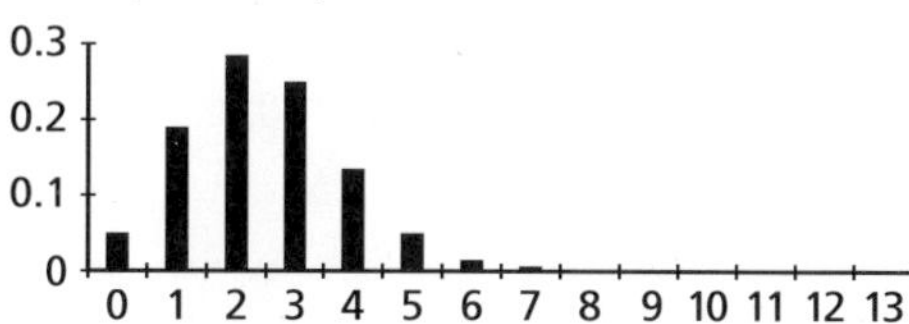


**c** The probability distribution of $X$ when $n = 20$

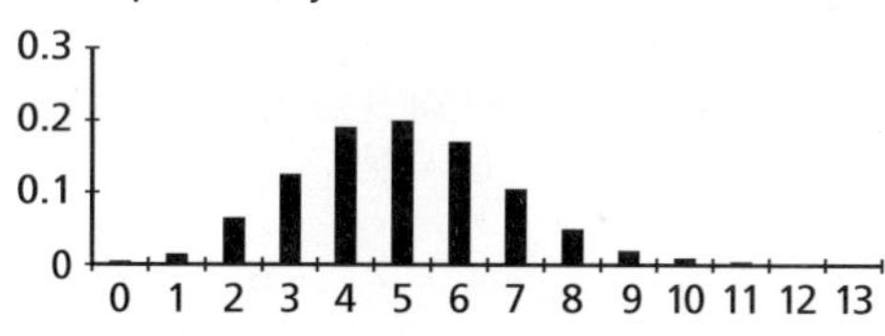


As $n$ increases the distribution becomes less skewed.

### Exercise 2.4A (page 48)

**1** **a** 0.5488 **b** 0.0988 **c** 0.0030

**2** **a** 0.0498, 0.1493, 0.2241, 0.2240, 0.1681
**b** 0.0041, 0.0225, 0.0618, 0.1133, 0.1558
**c** 0.0001, 0.0011, 0.0050, 0.0150, 0.0338

**3** 0.0000, 0.0001, 0.0004, 0.0018, 0.0053, 0.0127

**4** **a** 0.4497 **b** 0.1490 **c** 0.4013

**5** **a** 0.1003 **b** 0.5960 **c** 0.4040

**6** **a** 0.1353 **b** 0.2707 **c** 0.8571

**7** **a** 0.0821 **b** 1.2313

**8** **a**, **c** and **e** Poisson (see page 42)
**b** and **f** Binomial (see page 46)
**d** Not Binomial because number of trials is not fixed. Not Poisson because events are not taking place at an average rate over space or time.

### Exercise 2.4B (page 49)

**1** 0.0608, 0.1703, 0.2384, 0.2225, 0.1557, 0.0872

**2** **a** 0.3012 **b** 0.2169

**3** **a** $\dfrac{e^{-5} \times 5^k}{k!}$
**b** **(i)** 0.0337 **(ii)** 0.0843
**(iii)** 0.1755 **(iv)** 0.3840

**4** $X \sim \text{Poi}(7.5)$, $P(X \geq 6) = 0.7586$

**5** **a** $P(X = 0) = e^{-\mu} = 0.110\,803\,2$
$E(X) = V(X) = \mu = -\ln(0.110\,803\,2) \approx 2.2$
**b** 0.8194

**6** **a** **(i)** $P(W = 0) = P(X + Y = 0)$
$= P(X = 0) \times P(Y = 0)$
$= e^{-\mu_1} \times e^{-\mu_2} = e^{-(\mu_1 + \mu_2)}$
**(ii)** $P(W = 1) = P(X + Y = 1)$
$= P(X = 1)P(Y = 0) + P(X = 0)P(Y = 1)$
$= \mu_1 e^{-\mu_1} \times e^{-\mu_2} + e^{-\mu_1} \times \mu_2 e^{-\mu_2}$
$= (\mu_1 + \mu_2)e^{-(\mu_1 + \mu_2)}$

**b** **(i)** $P(W = 2) = \dfrac{e^{-(\mu_1 + \mu_2)}(\mu_1 + \mu_2)^2}{2!}$
**(ii)** $P(W = 3) = \dfrac{e^{-(\mu_1 + \mu_2)}(\mu_1 + \mu_2)^3}{3!}$

**c** $W \sim \text{Poi}(\mu_1 + \mu_2)$

**7** **a** The Poisson distribution, mean = 1

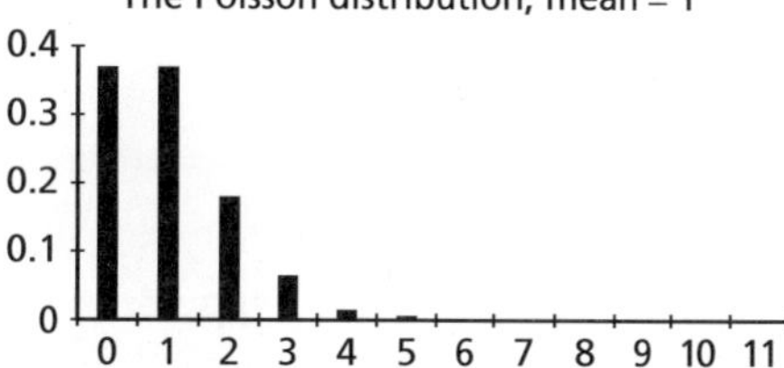


**b** The Poisson distribution, mean = 2

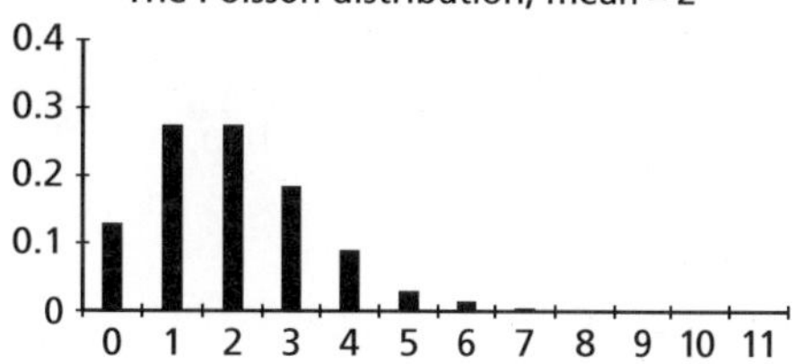


**c** The Poisson distribution, mean = 4

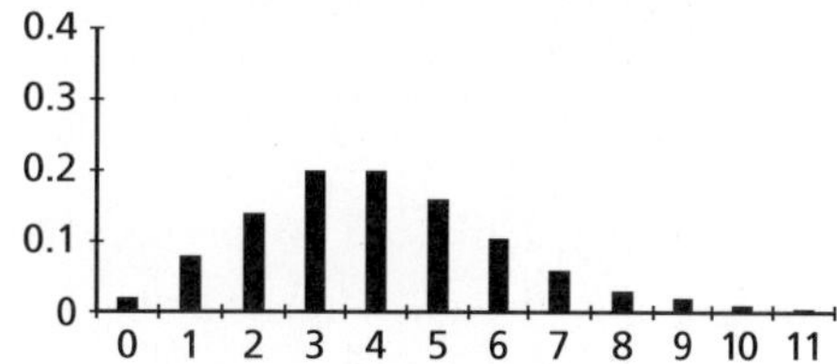


As the mean $\mu$ increases, the distribution becomes more symmetrical.

### Exercise 2.5B (page 50)

**1** **a** 0.6050, 0.3056, 0.0756
**b** 0.6065, 0.3033, 0.0758

**2** **a** 0.1766 **b** 0.1812

**3** **a** 0.2088 **b** 0.2083

**4** 0.4170
Selecting consecutive items may violate the assumption of independence. A random sample would be better.

**5** 0.2149
See page 46 for necessary conditions.

**6** 0.0296

### Exercise 2.6A (page 54)

**1** **a** $F(x) = x^3$, $0 \leq x \leq 1$
**b** **(i)** 0.216 **(ii)** 0.027 **(iii)** 0.189
**c** $\mu = \frac{3}{4}$, $\sigma^2 = \frac{3}{80}$

**2** **a** $f(x) = \begin{cases} \frac{1}{2}x & 0 \leq x \leq 2 \\ 0 & \text{otherwise} \end{cases}$

**b**

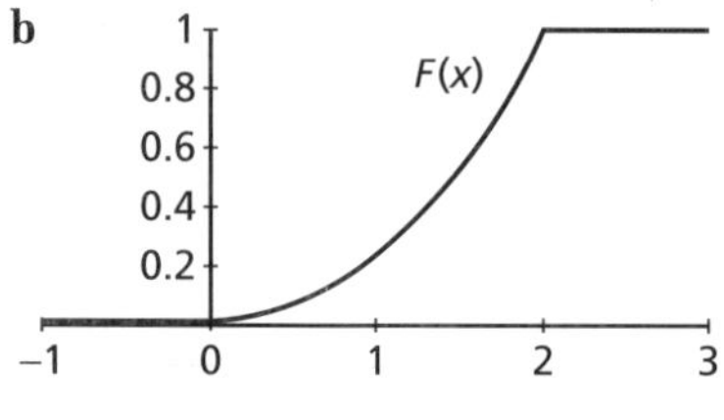


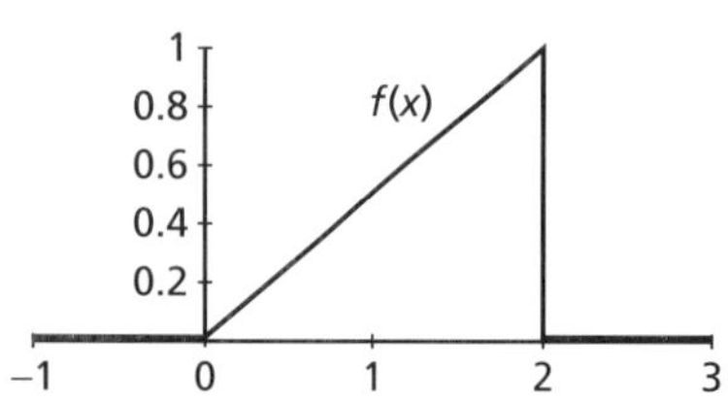


**c** $\mu = \frac{4}{3}, \sigma^2 = \frac{2}{9}$

**3** **a** $k = 5$ **b** $F(x) = \frac{1}{5}x, 0 \le x \le 5$

**c** **(i)** 0 **(ii)** 0.2 **(iii)** 0.2 **(iv)** 0.8

**4** **a** 0.2 **b** 0.2 **c** $\mu = 35, \sigma^2 = \frac{25}{3}$

**5** **a**

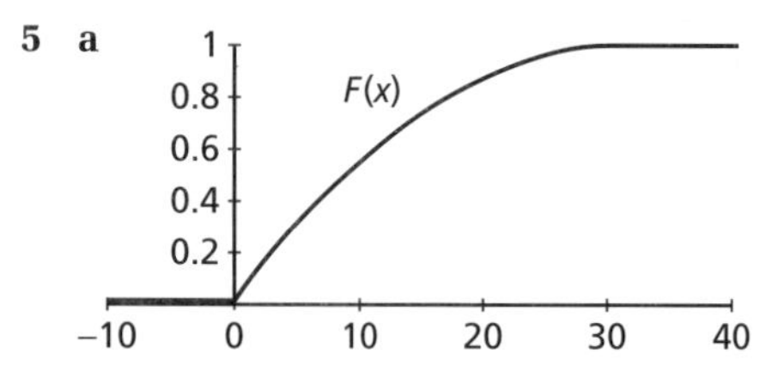


$F(x) = \frac{1}{900}(60x - x^2), 0 < x < 30$

**b** $\frac{1}{3}$ **c** $\mu = 10, \sigma^2 = 50$

**d** **(i)** $E(Y) = 20, V(Y) = 200$
**(ii)** $E(Y) = 15, V(Y) = 12.5$

**6** **a** $k = 6$ **b** $\mu = 0.5, \sigma = \frac{\sqrt{5}}{10} \approx 0.2236$

**c** $F(x) = 3x^2 - 2x^3, 0 \le x \le 1$

**d** **(i)** 0.028 **(ii)** 0.352 **(iii)** 0.34375

**7** **a** $\frac{5}{9}, \frac{1}{4}$ **b** $f(w) = \frac{32}{w^3}, w > 4$

### Exercise 2.6B (page 56)

**1** **a** $F(x) = 1 - \frac{100}{x}, x > 100$ **b** $m = 200$

**c** $\text{IQR} = 266\frac{2}{3}$

**2** **a** $E(X^r) = \int_0^1 2x^r(1 - x)dx, r = 1, 2, 3, \ldots$

**b** $E[(2X + 1)^2] = 4E(X^2) + 4E(X) + 1 = 3$

**3** **a** $k = \frac{6}{\pi^3}$

**b** $F(x) = \frac{1}{\pi^3}(3\pi x^2 - 2x^3), 0 < x < \pi$

**c** Length > 10 cm $\Leftrightarrow X > \frac{\pi}{3}$ radians

$P\left(X > \frac{\pi}{3}\right) = \frac{20}{27} \approx 0.7407$

**4** **a** $k = -\frac{1}{4500}$ **b** $E(T) = 15\,°C$

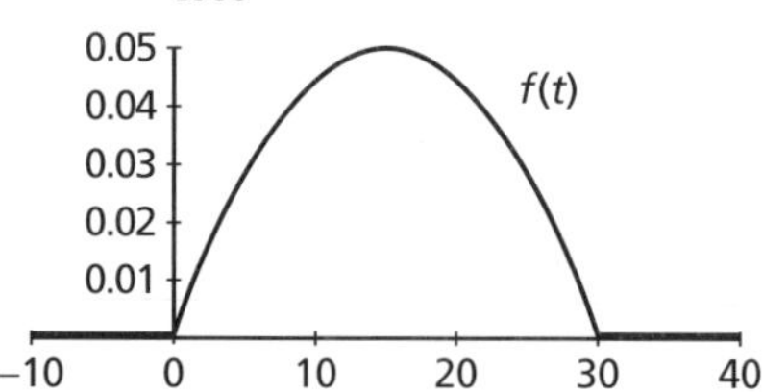


**c** $V(T) = 45\,°C^2$

**d** $E(T) = 59\,°F, V(T) = 145.8\,°F^2$

**5** $\mu = \frac{a + b}{2}, \sigma^2 = \frac{(b - a)^2}{12}$

**6** **a**

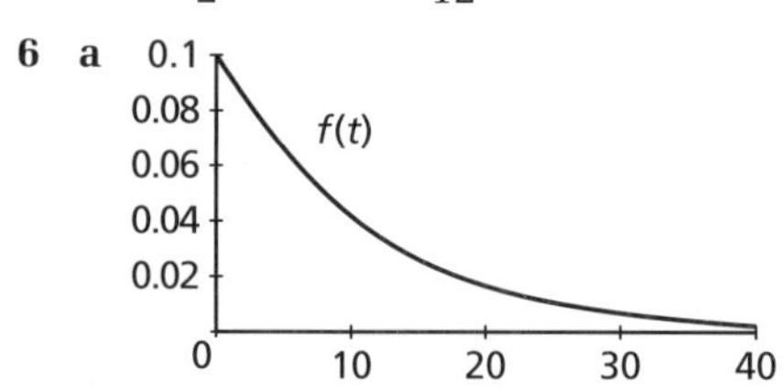


**b** **(i)** 0.1813 **(ii)** 0.2231

**7** $\mu = \frac{1}{k}, \sigma = \frac{1}{k}$

### Exercise 2.7 (page 60)

All answers are derived from 2 d.p. values of $Z$

**1** **a** 0.5000 **b** 0.8413 **c** 0.9772
**d** 0.9162 **e** 0.0838 **f** 0.0170

**2** **a** 0 **b** 0.5000 **c** 0.1429
**d** 0.1587 **e** 0.0158 **f** 0.1429
**g** 0.8413 **h** 0.9842 **i** 0.6826
**j** 0.9544 **k** 0.9974 **l** 0.9500

**3** **a** 0.8413 **b** 0.1587 **c** 0.1587
**d** 0.8413 **e** 0.6826 **f** 0.9544
**g** 0.4514 **h** 0.7333 **i** 0.6612

**4** **a** 2.17 **b** 0.31 **c** 0.67
**d** −1.96 **e** −0.84 **f** −0.18

**5** **a** **(i)** 77.28 **(ii)** 71.44
**b** **(i)** 337.5 **(ii)** 332.5
**c** 36.1, 40.9, 45.7; quartiles

**6** 0.0968

**7** 1516.4 g

**8** 62 g

**9** **a** 1.4 **b** 64 **c** $New = \frac{2}{3}(old + 20)$
**d** **(i)** 50 **(ii)** 36 **(iii)** 68

**10** $P(X < 60) = 0.8, P(X < 40) = 0.1$

Solve: $\mu + \Phi^{-1}(0.8)\sigma = 60$
$\mu + \Phi^{-1}(0.1)\sigma = 40$

$\mu \approx 52.1$ mph, $\sigma \approx 9.4$ mph

### Exercise 2.8B (page 62)

**1** 0.7967

**2** **a** 0.7088 **b** 0.7088 **c** 0.6603
**d** 0.4070

**3** 0.2206

4 a $E\left(\frac{X-\mu}{\sigma}\right) = \frac{1}{\sigma}E(X) - \frac{\mu}{\sigma} = 0$

b $V\left(\frac{X-\mu}{\sigma}\right) = \frac{1}{\sigma^2}V(X) = 1$

5 a 0.0594
Weights of man and woman are independent.

b 163 kg

6 a 0.7257

b $T = M_1 + \cdots + M_{10} + F_1 + \cdots + F_8$
$T \sim N(62.4, 2.32)$

7 a N(90, 144) b N(3240, 5184)

c 0.9996

8 0.1587

**Exercise 2.9B (page 65)**

1 a 0.5881

b (i) 0.5871 (ii) 0.5000

2 a (i) 0.5398 (ii) 0.5000

b (i) 0.5120 (ii) 0.5000

c The discrepancy becomes less.

3 a $X \sim \text{Bin}(30, 0.5)$
$\mu = 15, \sigma = \sqrt{7.5} \approx 2.74$

b (i) 0.8186 (ii) 0.8997
(iii) 0.5714 (iv) 0.7088

4 0.9382

5 0.7852

6 a $X \sim \text{Bin}(230, 0.8)$ b 0.0033

c People sometimes travel in groups

7 0.0668

**Chapter 2 Review Exercise (page 73)**

1

| x | 2 | 3 | 4 | 5 | 6 |
|---|---|---|---|---|---|
| p(x) | $\frac{1}{6}$ | $\frac{1}{6}$ | $\frac{1}{6}$ | $\frac{1}{6}$ | $\frac{1}{3}$ |

$\mu = \frac{13}{3}, \sigma^2 = \frac{20}{9}$

2 a $E(X+Y) = 30, V(X+Y) = 13$

b $E(X-Y) = -10, V(X-Y) = 13$

c $E(5X-2Y) = 10, V(5X-2Y) = 136$

d $E(2X+Y-15) = 25$
$V(2X+Y-15) = 25$

3 a 0.2269 b 0.0250 c 0.6471

d 0.3529

4 a 0.0155 b $\mu = 2.4, \sigma = \sqrt{1.92} \approx 1.4$

5 0.4966, 0.3476, 0.1217, 0.0284

6 0.1736

7 a 0.1353 b 0.2707 c 0.6767

8 a $k = -\frac{3}{32}$

b $F(x) = \frac{1}{32}\left(7 - 15x + 9x^2 - x^3\right), 1 < x < 5$

c $\frac{11}{16}$

9 a 0.9082 b 0.0146 c 1.63

d −0.44

10 a 0.0228 b 0.3085 c 0.8664

11 $691.5 \approx 692$

12 0.3745

13 0.3015

14 0.0015

# CHAPTER 3

**Exercise 3.1 (page 77)**

1 See Populations and Samples page 75.

2 See Populations and Samples page 76.

3 See Populations and Samples page 76.

4 a Individual first-year girls

b All first-year girls at this school

c A list of all first-year girls' names

5 a Individual parents/guardians

b The parents/guardians of all children attending this school

c The school office will have a list of all those entitled to vote in School Board elections

6 a Calculate the average number of chairs in a sample of classrooms.

b (i) Classrooms
(ii) All classrooms in the school
(iii) A list of all classrooms

7 a Inspect the date stamps in a sample of books.

b (i) Books
(ii) All books in the library
(iii) The library's catalogue

8 a Choose a sample of unrelated pupils and ask them how many TVs they have.
Calculate the sample mean.
(i) Pupils' families
(ii) All pupils' families at the school
(iii) The school roll, but make sure sample does not include siblings

b Choose a sample of households and send them a questionnaire.
Calculate the proportion of households in the sample with more than one car.
(i) Households
(ii) All households in town
(iii) The Council Tax list.

c Divide the beach into sections all the same size and calculate the mean amount of seaweed in a sample of these sections. Multiply this sample mean by the number of sections.
(i) Numbered sections
(ii) All sections, i.e. the whole beach
(iii) A list of numbers.

**Exercise 3.2 (page 78)**

Answers will vary. On one occasion the following results were obtained:

| Simple random sampling | Sample mean | Non-random sampling |
|---:|:---:|:---|
| | 0.5 | |
| • | 0.6 | |
| • | 0.7 | |
| • • • | 0.8 | |
| • • • | 0.9 | |
| • • | 1.0 | • |
| • • • | 1.1 | • |
| • • • • | 1.2 | • |
| • • | 1.3 | • • • |
| • | 1.4 | • • • • |
| | 1.5 | • • • |
| | 1.6 | • • |
| | 1.7 | • • • |
| | 1.8 | • |
| | 1.9 | |
| | 2.0 | • |

The true population mean diameter is 1.0 cm.

There are $\binom{60}{5} = 5\,461\,512$ simple random samples of size 5 and they are all equally likely.

**Exercise 3.3 (page 81)**

**1** **a** $9.517\,025 \approx 9.52$ **b** 1.5 **c** 0.3745

**2** **a** 0.2514 **b** 2.5 **c** 0.0228

**3** **a** 0.1587 **b** $100/\sqrt{6} \approx 40.825$ **c** 0.0071

**4** **a** 0.0796 **b** $\bar{X} \sim \mathrm{N}\left(\mu, \frac{\sigma^2}{100}\right)$, 0.6826 **c** $n \geq 384.16$, say $n = 385$

**Exercise 3.4 (page 84)**

**1** **a**

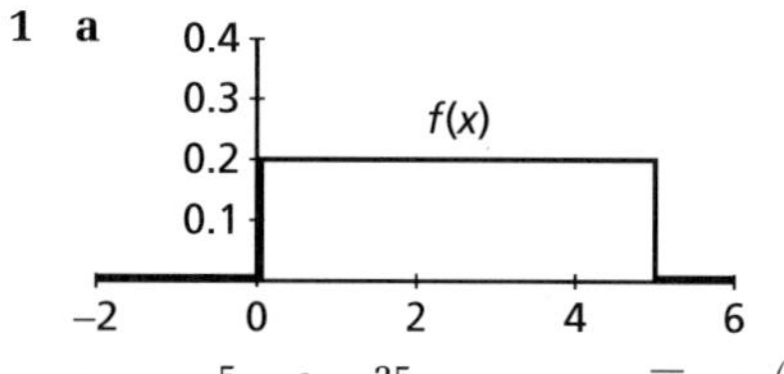


**b** $\mu = \frac{5}{2}, \sigma^2 = \frac{25}{12}$ **c** $\bar{X} \sim \mathrm{N}\left(\frac{5}{2}, \frac{1}{12}\right)$

**d** **(i)** 0.4 **(ii)** 0.0418

**2** **a** $\bar{X} \sim \mathrm{N}(10, 2)$

**b** **(i)** 0.0793 **(ii)** 0.2389

**3** **a** $\mathrm{E}(X_i) = 10$, $\mathrm{V}(X_i) = 5$

**b** $T \sim \mathrm{N}(300, 150)$ **c** 0.5160

**d** 0.5160

**4** **a** $\mu = 10, \sigma^2 = 50$ **b** $\bar{X} \sim \mathrm{N}(10, 2)$

**c** So that we may assume independence

**d** **(i)** $\frac{371}{900} \approx 0.4122$ **(ii)** 0.0170

**5** **a**

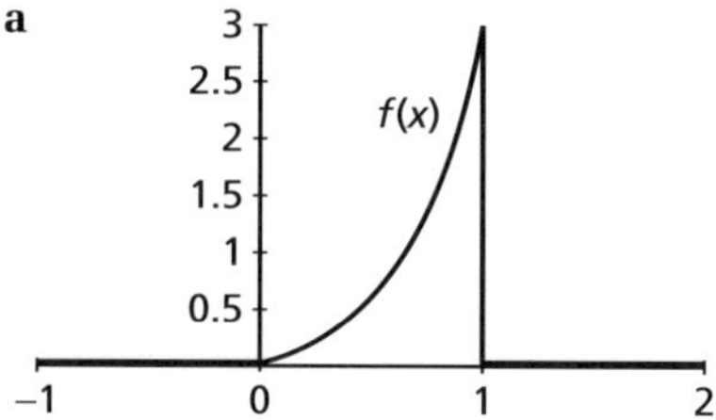


**b** $\mu = 0.75, \sigma^2 = \frac{3}{80}$

**c** $\bar{X} \sim \mathrm{N}\left(0.75, \frac{1}{960}\right)$

**d** So that we may assume independence

**e** **(i)** 0.216 **(ii)** 0.0000

**6** **a** $\mathrm{N}\left(175, \frac{400}{30}\right)$ **b** 0.0853

**7** **a** $\mu = 0.4, \sigma^2 = 0.24$

**b** $T \sim \mathrm{N}(10, 6)$, 0.9878

**c** 0.9878, same

**Exercise 3.5A (page 88)**

**1** **a** **(i)** 161.4 **(ii)** 3.3541

**b** (154.83, 167.97)

95% of such intervals will capture the true mean height so the population mean height for women is very likely to be between 154.83 and 167.97 cm.

**2** **a** **(i)** 101.625 **(ii)** 7.0711

**b** (89.99, 113.26) **c** (87.77, 115.48)

The higher the confidence level, the wider the interval.

**3** (25.56, 26.00)

95% of such CIs will capture the true mean contents of this brand of crisps so the population mean is very likely to be between 25.56 and 26.00 g.

**4** (59.9, 73.0) 95% of such CIs will capture the true mean weight of male students so the population mean is very likely to be between 59.9 and 73.0 kg.

**5** **a**

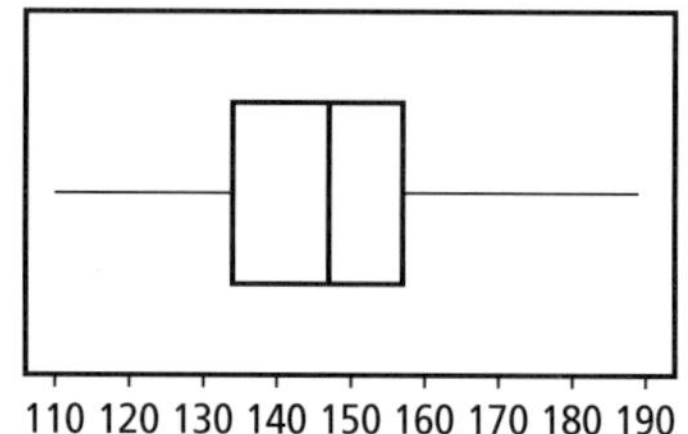


**b** (137.09, 157.33)

95% of such CIs will capture the true mean potassium level in men so the population mean is very likely to be between 137.09 and 157.33 Bq/kg.

### Exercise 3.5B (page 89)

**1 a** $N\left(\mu, \dfrac{0.08^2}{65}\right)$

**b** (0.301, 0.339)

95% of such CIs will capture the true mean shell thickness so the population mean is very likely to be between 0.301 and 0.339 mm.

**2 a**

| Stem | Leaf |
|---|---|
| 3 | 0 1 3 |
| 4 | 0 5 7 9 |
| 5 | 1 3 5 8 9 |
| 6 | 1 3 5 6 9 |
| 7 | 2 4 9 |
| 8 | 0 5 8 |
| 9 | 3 |

$n = 24$ 9 | 3 represents 93 marks

These marks have a symmetrical distribution similar to the Normal.

**b** 60.25 **c** $\overline{X} \sim N\left(\mu, \dfrac{20^2}{24}\right)$

**d (i)** (53.55, 66.95) **(ii)** (52.25, 68.25)

**(iii)** (49.72, 70.78)

The higher the confidence level, the wider the interval.

**3 a** (146.82, 149.18)

**b** It would be unwise to use them because 150 lies outwith the CI.

**4 a**

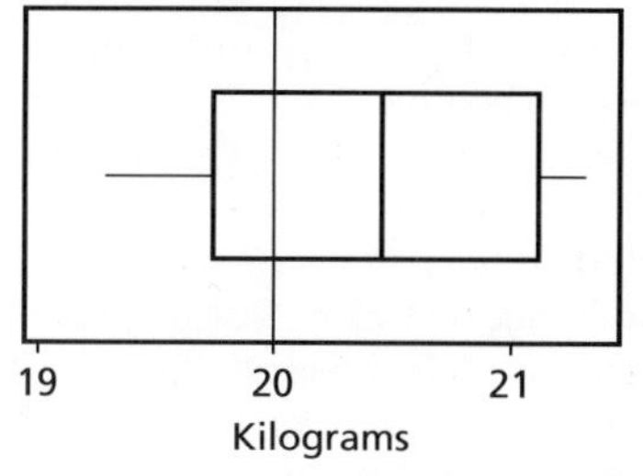


The castings weights seem symmetrically distributed and more than half of them are greater than 20 kg.

**b** (19.9, 20.8)

**c** Since 20 kg lies in the CI, the foundry's claim is justified.

**5 a**

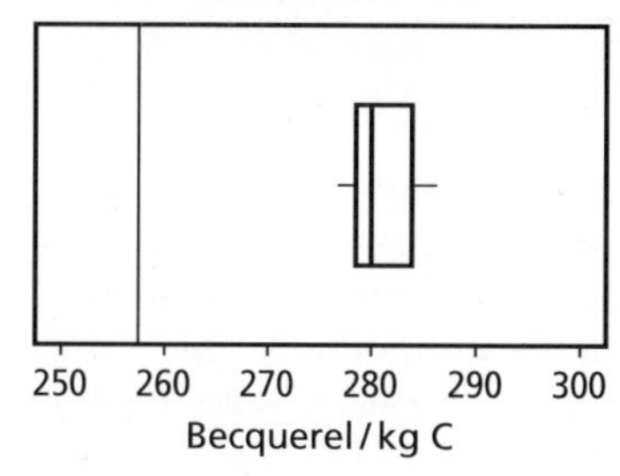


**b** (279.43, 282.57)

**c** The level of radiocarbon surrounding the plant is much higher than the global environmental level. The value 257 Bq/kg lies well below the lower limit of the interval, confirming the boxplot.

**6 a**

| Stem | Leaf |
|---|---|
| 0 | 8 |
| 1 | 0 |
| 1 | 2 3 |
| 1 | 4 4 4 5 |
| 1 | 7 |
| 1 | 8 8 9 9 |
| 2 | 0 0 1 1 1 |
| 2 | 3 3 |
| 2 | 4 4 5 |
| 2 | |
| 2 | 8 9 |

$n = 25$ 2 | 8 represents 2.8

The shape of the distribution is reasonably symmetrical.

**b** (1.684, 2.076)

95% of such CIs will capture the true mean amplitude so the population mean is very likely to be between 1.684 and 2.076.

**7 a** (0.504, 1.092)

**b**

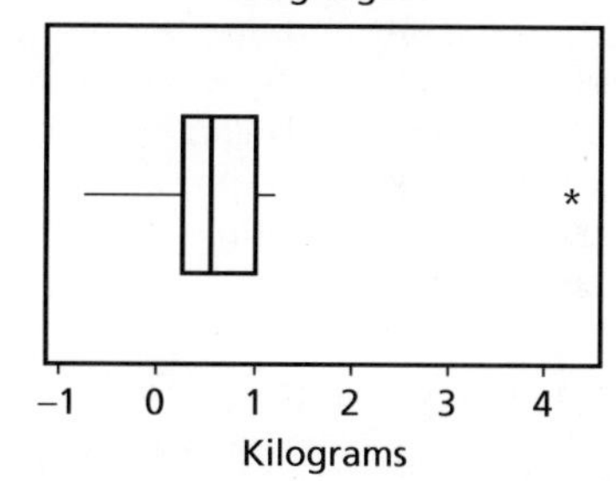


**c** (0.173, 0.788)

Both limits of the CI are lower.

**8** $1.96 \times \dfrac{4}{\sqrt{n}} = 1 \Rightarrow n \approx 62.$

### Exercise 3.6B (page 92)

**1 a (i)** 0.64 **(ii)** 0.048

**b** (0.54592, 0.73408)

**c** The interval does not contain 0.5 so the coin seems to be producing too many Heads.

**2 a (i)** 0.3 **(ii)** 0.0837

**b** (0.136, 0.464)

95% of such CIs will capture the true proportion of pupils who travel to school by car so the population proportion is very likely to be between 0.136 and 0.464.

**3** (0.199, 0.241)
What voters say they will do before polling day and what they actually do on the day may be different.

**4** **a** (0.437, 0.511) **b** (0.430, 0.518)
**c** (0.416, 0.532)
The higher the level of confidence is, the wider the interval.

**5** **a** (0.352, 0.548)
**b** Females (0.271, 0.516)
Males (0.382, 0.695)
Since the intervals overlap there is no evidence that different proportions of females and males are physically active.

**6** **a** 95% CI is (0.060, 0.125)
**b** (0.800, 0.883)
The opinions of patients are very different from the rating given by the consultant.

**7** **a** (0.435, 0.837)
**b** The interval contains 0.5 so twins may be equally likely to hold similar or dissimilar views.

### Exercise 3.7 (page 96)

**1** See Stratified random sampling page 94.

**2** Preferences in TV viewing may vary with age and sex. Obtain a list of the pupils in each of the 12 strata and number pupils consecutively from 1 to $N_s$ where $N_s$ is the number in the stratum. Using $n_s$ distinct random numbers select $n_s$ pupils from the numbered list for the stratum, where

$$n_s = 0.1 \times N_s$$

Repeat this for each stratum separately.

**3** Answers will vary. On one occasion the following multiple boxplot was obtained.

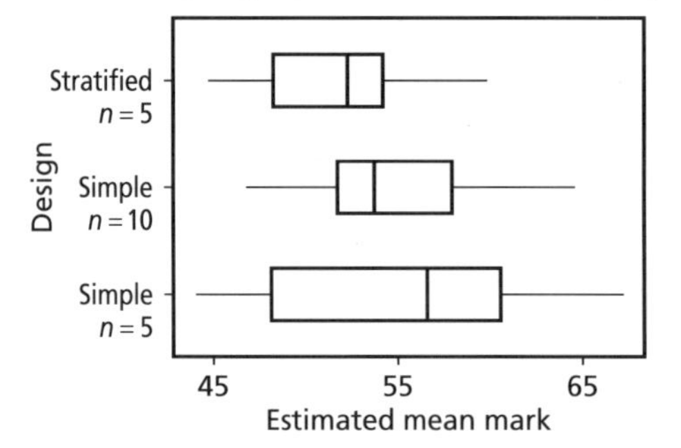


The population median mark is 54 and each of the three sampling methods produces estimates which are centred close to this value. However, the interquartile range (IQR) of the estimates produced by stratified samples of size 5 and simple random samples of size 10 are similar and both are less than the IQR for estimates produced by simple random samples of size 5. The smaller the IQR is, the more precision we can expect from the estimates. The stratified random samples of size 5 are efficient because they seem to give similar precision to simple random samples of size 10 but for less cost (smaller sample size).

### Exercise 3.8 (page 98)

**1** See Cluster sampling page 96.

**2** See Cluster sampling page 97.

**3** Answers will vary.

| Class | $\hat{p}$ | Class | $\hat{p}$ |
|---|---|---|---|
| 1 | 0.091 | 13 | 0.031 |
| 2 | 0.152 | 14 | 0.121 |
| 3 | 0.091 | 15 | 0.129 |
| 4 | 0.061 | 16 | 0.094 |
| 5 | 0.156 | 17 | 0.107 |
| 6 | 0.121 | 18 | 0.040 |
| 7 | 0.063 | 19 | 0.136 |
| 8 | 0.121 | 20 | 0.240 |
| 9 | 0.182 | 21 | 0.056 |
| 10 | 0.152 | 22 | 0.136 |
| 11 | 0.125 | 23 | 0.105 |
| 12 | 0.063 | 24 | 0.095 |

### Exercise 3.9 (page 100)

**1** **a** Convenience
**b** Stratified random sampling
**c** Cluster
**d** Quota
**e** Systematic
**f** Simple random sampling
**g** Convenience (self-selected)
Refer to discussion on pages 77, 94, 96, 98.

**2** See Further sampling methods page 98.

**3** **a** Number all voters from 1 to 100 000. Choose a number at random in the range 1 to 50 inclusive; select the voter with this number and then every 50th voter on the list thereafter.
**b** Estimate is $\hat{p} = \dfrac{291}{560} = 0.520$
95% CI is (0.478, 0.561)
Caution should be exercised when interpreting a CI constructed from a systematic sample.
**c** Although questionnaires were sent to 2000 voters only 28% of these replied. The council should be concerned about what the 72% who did not reply think about council services.

## Chapter 3 Review Exercise (page 109)

1 See Populations and samples page 75.

2 a 24 b 2 c $\overline{X} \sim N(20, 4)$

d (i) 0.8413 (ii) 0.9772

3 a $E(X_i) = 9$, $V(X_i) = 3.6$

$T \sim N(225, 90)$

b $P(\overline{X} < 10) = P(T < 250) = 0.9951$

4 a $\overline{X} \sim N(0.4, 0.01)$ b 0.8413

5 See relevant section(s) in Chapter 3.

6 See relevant section(s) in Chapter 3.

7 a

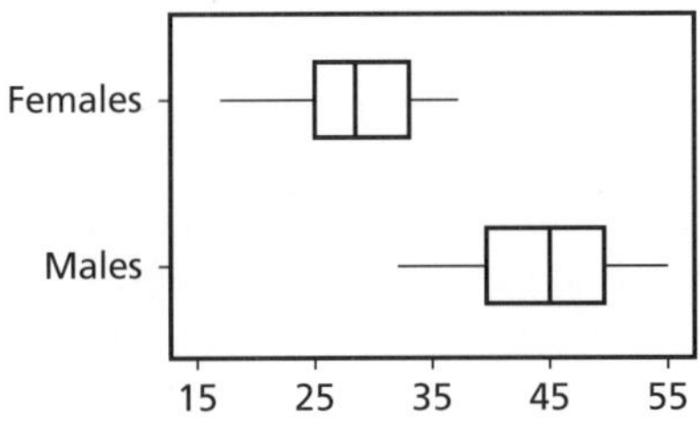


b (i) 44.92 (ii) 28.50

c (i) 1.225 (ii) 1.021

d (i) (42.52, 47.32) (ii) (26.50, 30.50)

e The CIs do not overlap. There is a very significant difference between the grip strength of male and female students. This can also be seen in the boxplot.

8 a 0.35 b 0.0477

c (0.26, 0.44).
95% of such CIs will capture the true proportion of books borrowed more than once, so the population proportion is very likely to be between 0.26 and 0.44.

# CHAPTER 4

## Exercise 4.1 (page 113)

1 Let $p$ be the population proportion of Scottish adults who brush their teeth at least twice daily.

$H_0$: $p = 0.74$ $H_1$: $p > 0.74$

2 Let $\mu$ grams be the population mean weight of a packet of crisps.

$H_0$: $\mu = 25$ $H_1$: $\mu < 25$

3 Let $\mu$ mm be the population mean cross-sectional diameter of a metal rod.

$H_0$: $\mu = 5$ $H_1$: $\mu > 5$

4 Let $\mu$ be the population mean score of a Secondary 2 pupil.

$H_0$: $\mu = 22$ $H_1$: $\mu \neq 22$

5 Let $\mu_D$ be the population mean change in fitness score for a female who follows the exercise programme.

$H_0$: $\mu_D = 0$ $H_1$: $\mu_D > 0$

6 Let $\mu$ be the population mean IQ score of prisoners convicted of theft.

$H_0$: $\mu = 100$ $H_1$: $\mu \neq 100$

7 Let $p$ be the population proportion of defective items produced by the upgraded facility.

$H_0$: $p = 0.045$ $H_1$: $p < 0.045$

8 Let $\mu_D$ be the population mean change in the annual number of hospital admissions for a teenager who has been given asthma training.

$H_0$: $\mu_D = 0$ $H_1$: $\mu_D < 0$

## Exercise 4.2 (page 116)

1 a In general, the boxplot suggests experts seem to estimate a bit low. Normality seems acceptable (although there is one very low outlier).

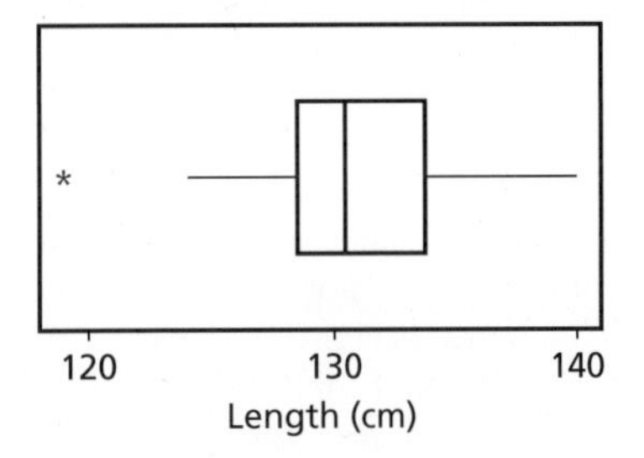


b Let $\mu$ cm be the population mean estimate by an expert. Assume the population standard deviation is 5 cm.

$H_0$: $\mu = 135$ vs $H_1$: $\mu \neq 135$

$$\overline{x} = 130.7,\ z = \frac{130.7 - 135.0}{5/\sqrt{20}} = -3.85$$

2 a In general, the boxplot suggests blood glucose levels are a bit higher than normal. Normality seems acceptable.

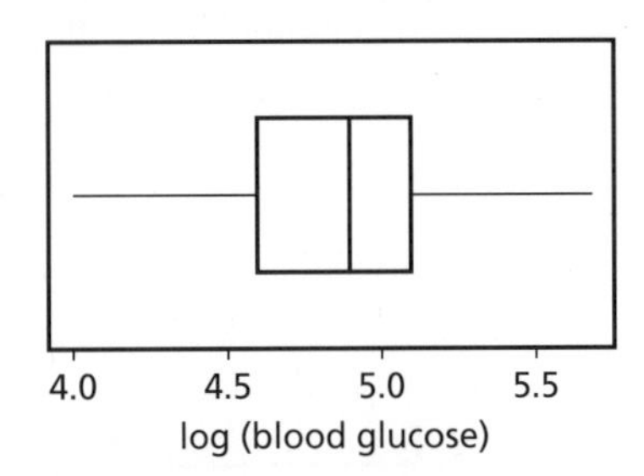


b Let $\mu$ cm be the population mean log blood glucose level for male diabetics. Assume the population standard deviation is 0.5.

$H_0$: $\mu = 4.7$ vs $H_1$: $\mu > 4.7$

$$\overline{x} = 4.868,\ z = \frac{4.868 - 4.700}{0.5/\sqrt{25}} = 1.68$$

**3 a** Let $\mu$ cm be the population mean length of baby boys in Glasgow. Assume the population standard deviation is 2.02 cm.

$H_0: \mu = 51.09$ vs $H_1: \mu \neq 51.09$

$\bar{x} = 51.97,\ z = \dfrac{51.97 - 51.09}{2.02/\sqrt{44}} = 2.89$

**b** Let $\mu$ cm be the population mean length of baby girls in Glasgow. Assume the population standard deviation is 1.88 cm.

$H_0: \mu = 50.21$ vs $H_1: \mu \neq 50.21$

$\bar{x} = 51.11,\ z = \dfrac{51.11 - 50.21}{1.88/\sqrt{41}} = 3.07$

**Exercise 4.3 (page 118)**

**1** $p = 2\Phi(-3.85) = 2[1 - \Phi(3.85)] < 0.0002$
This value of $p$ is very low, so the sample data are not consistent with $H_0$. The population mean estimate of length is significantly different from 135 cm, which is the true value.

**2** $p = P(Z \geq 1.68 \mid H_0) = 1 - \Phi(1.68) = 0.0465$
This value of $p$ is low ($< 0.05$), so the sample data are not consistent with $H_0$. The population mean log blood glucose level of male diabetics is significantly greater than 4.7.

**3 a** $p = 2[1 - \Phi(2.89)] = 0.0038$
This value of $p$ is very low, so the sample data are not consistent with $H_0$. The population mean length of new-born baby boys in Glasgow is significantly different from 51.09 cm.

**b** $p = 2[1 - \Phi(3.07)] = 0.0022$
This value of $p$ is very low, so the sample data are not consistent with $H_0$. The population mean length of new-born baby girls in Glasgow is significantly different from 50.21 cm.

**4 a** The multiple boxplot below suggests that, on average, both sixth-year boys and girls complete the task in less than 75 seconds. The Normality assumption seems a bit unlikely here, since both data sets appear to be positively skewed.

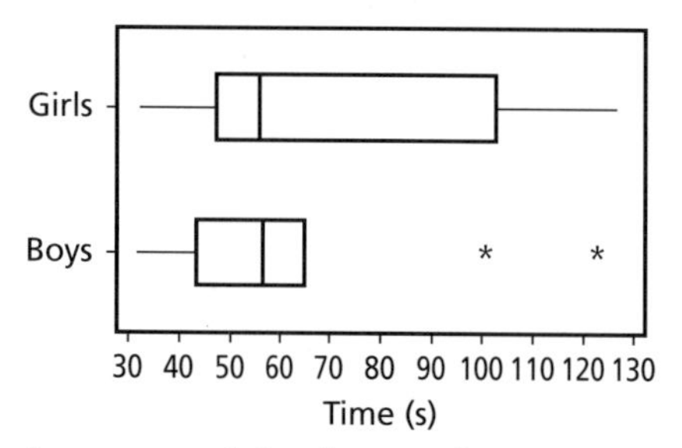


**b** Let $\mu$ seconds be the population mean time required to complete the sorting task. Assume the population standard deviation is 30 seconds.

$H_0: \mu = 75$ vs $H_1: \mu > 75$

For boys,

$\bar{x} = 63.10,\ z = \dfrac{63.10 - 75.00}{30/\sqrt{10}} = -1.25$

$p = P(Z \geq -1.25 \mid H_0) = \Phi(1.25) = 0.8944$

This value of $p$ is high, so there is no reason to reject $H_0$ in favour of $H_1$. The population mean time required to complete the sorting task is not significantly greater than 75 seconds.

For girls,

$\bar{x} = 68.30,\ z = \dfrac{68.30 - 75.00}{30/\sqrt{10}} = -0.71$

$p = P(Z \geq -0.71 \mid H_0) = \Phi(0.71) = 0.7611$

This value of $p$ is high, so there is no reason to reject $H_0$ in favour of $H_1$. The population mean time required to complete the sorting task is not significantly greater than 75 seconds.

**Exercise 4.4 (page 121)**

**1 a** $CR = \{z: z < -z_{0.025}\ or\ z > z_{0.025}\}$
$= \{z: z < -1.96\ or\ z > 1.96\}$
$z = -3.85$ lies in CR, so reject $H_0$ in favour of $H_1$.

**b** Conclusions are the same.

**2 a** $CR = \{z: z > z_{0.01}\}$
$= \{z: z > 2.33\}$
$z = 1.68$ is not in CR, so do not reject $H_0$.

**b** Conclusions are not the same. Although $p = 0.0465$ is small, it is not smaller than 0.01.

**3 a** $CR = \{z: z > z_{0.05}\}$
$= \{z: z > 1.64\}$
For boys, $z = -1.25$. For girls, $z = -0.71$. Neither value lies in CR, so do not reject $H_0$ for boys or girls.

**b** Conclusions are the same.

**4** Let $\mu$ grams be the population mean weight of a packet of crisps. Assume the population standard deviation is 1 gram.

$H_0: \mu = 25$ vs $H_1: \mu > 25$

$\bar{x} = 26.082,\ z = \dfrac{26.082 - 25.000}{1/\sqrt{50}} = 7.65$

$CR = \{z: z > z_{0.001}\} = \{z: z > 3.09\}$

$z$ lies in CR, so reject $H_0$ in favour of $H_1$. The population mean weight of packets of these crisps is significantly greater than 25 grams.

**Exercise 4.5B (page 123)**

**1** The multiple boxplot below suggests that, on average, sixth-year boys and girls both improve by more than 15 seconds. The

Normality assumption seems reasonable since both plots are fairly symmetric.

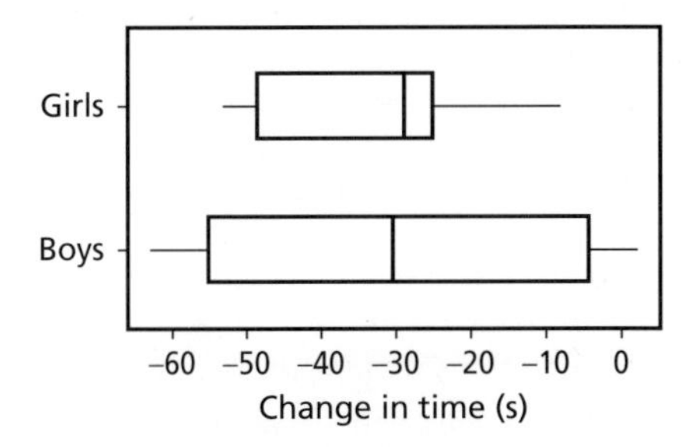


Let $\mu_D$ cm be the population mean change in the time required to complete the sorting task. Assume the population standard deviation is 20 seconds.

$H_0$: $\mu_D = -15$ vs $H_1$: $\mu_D < -15$

For girls,

$$\bar{d} = -34.6,\ z = \frac{-34.6 - (-15.0)}{20/\sqrt{10}} = -3.10$$

$$p = P(Z \le -3.10 \mid H_0) = 1 - \Phi(3.10) = 0.0010$$

This value of $p$ is low, so reject $H_0$ in favour of $H_1$. The population mean improvement for sixth-year girls is significantly better than 15 seconds.

For boys,

$$\bar{d} = -30.40,\ z = \frac{-30.40 - (-15)}{20/\sqrt{10}} = -2.43$$

$$p = P(Z \le -2.43 \mid H_0) = 1 - \Phi(2.43) = 0.0075$$

This value of $p$ is low, so reject $H_0$ in favour of $H_1$. The population mean improvement for sixth-year boys is significantly better than 15 seconds.

**2** **a** The boxplot below shows that there was a deterioration in the lung function of just over half this sample of children, which does not lend much support to the study hypothesis. The very low outlier and the general skewness of the plot cast doubt on the Normality assumption.

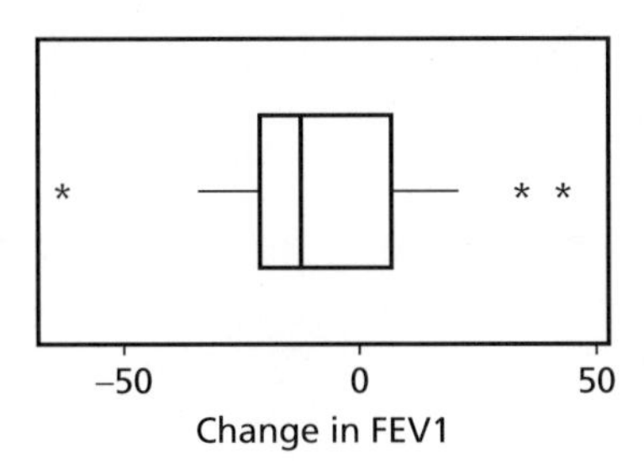


**b** Let $\mu_D$% be the population mean reduction in FEV1. Assume the population standard deviation is 25%.

$H_0$: $\mu_D = 0$ vs $H_1$: $\mu_D < 0$

$$\bar{d} = -7.11,\ z = \frac{-7.11 - 0}{25/\sqrt{18}} = -1.21$$

$$p = P(Z \le -1.21 \mid H_0) = 1 - \Phi(1.21) = 0.1131$$

This value of $p$ is not very low, so do not reject $H_0$. In this population, there is no significant mean deterioration in FEV1.

**3** Let $\mu_D$ kg be the population mean difference in determinations of FFM by the test and reference methods. Assume the population standard deviation is 3 kg.

$H_0$: $\mu_D = 0$ vs $H_1$: $\mu_D \ne 0$

Significance level is chosen to be 0.01, so

$$CR = \{z: z < -z_{0.005} \text{ or } z > z_{0.005}\} = \{z: z < -2.58 \text{ or } z > 2.58\}$$

$$\bar{d} = 2.72,\ z = \frac{2.72 - 0}{3/\sqrt{57}} = 6.85$$

$z$ lies in CR, so reject $H_0$ in favour of $H_1$. On average, FFM by the test method is significantly different from FFM by the reference method.

**Exercise 4.6B (page 126)**

**1** **a** (i) For boys, 95% confidence interval for $\mu$ is $\bar{x} \pm 1.96\frac{\sigma}{\sqrt{n}} = 51.97 \pm 1.96\frac{2.02}{\sqrt{44}}$ or (51.37, 52.57)

(ii) For girls, 95% confidence interval for $\mu$ is $\bar{x} \pm 1.96\frac{\sigma}{\sqrt{n}} = 51.11 \pm 1.96\frac{1.88}{\sqrt{41}}$ or (50.53, 51.69)

**b** In both cases, $\mu_0$ does not lie in the 95% confidence interval. Previously found that $p < 0.05$, so conclusions agree.

**2** **a** A 99% confidence interval for $\mu_D$ is

$$\bar{d} \pm 2.58\frac{\sigma_D}{\sqrt{n}} = 2.72 \pm 2.58\frac{3}{\sqrt{57}}$$

or (1.69, 3.75)

**b** The value 0 does not lie in the 99% confidence interval. Previously found, when testing $H_0$: $\mu_D = 0$, that $p < 0.01$, so conclusions agree.

**3** **a** Let $\mu$ mm be the population mean guess of length. Assume the population standard deviation is 27.6 mm.

$H_0$: $\mu = 138$ vs $H_1$: $\mu \ne 138$

Significance level is 0.05, so

$$CR = \{z: z < -1.96 \text{ or } z > 1.96\}$$

$$\bar{x} = 141.25,\ z = \frac{141.25 - 138.00}{27.6/\sqrt{40}} = 0.74$$

$z$ does not lie in CR, so do not reject $H_0$. The population mean guess is not significantly different from the true value, 138 mm.

**b** A 95% confidence interval for $\mu$ is

$$\bar{x} \pm 1.96\frac{\sigma}{\sqrt{n}} = 141.25 \pm 1.96\frac{27.6}{\sqrt{40}}$$

or (132.7, 149.8)

The true line length, 138 mm, lies in this confidence interval, so it is a plausible value for the population mean guess of the length.

**c** The two conclusions agree.

**Chapter 4 Review Exercise (page 130)**

**1 a** The boxplot shows that the sample median experimental value is close to 792.5, the 'true' speed. Normality seems reasonable, though perhaps the data are a little negatively skewed.

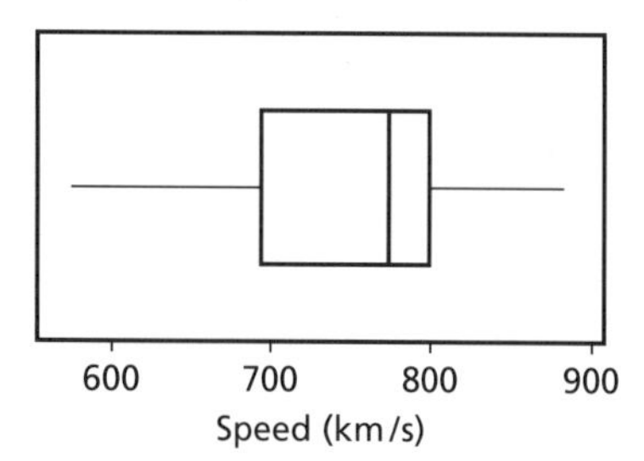


**b** Let $\mu$ km/s be the population mean speed determined by Michelson's experiments. Assume the population standard deviation is 5 km/s. Then a 95% confidence interval for $\mu$ is

$$\bar{x} \pm 1.96\frac{\sigma}{\sqrt{n}} = 742.8 \pm 1.96\frac{90}{\sqrt{22}}$$

or (705.2, 780.4)

The 'true' value, 792.5, does not lie in this confidence interval. It looks as though Michelson's experiments under-estimated the speed of light.

**c** $H_0$: $\mu = 792.5$ vs $H_1$: $\mu \neq 792.5$

Significance level is 0.05, so

CR = $\{z: z < -1.96 \text{ or } z > 1.96\}$

$$\bar{x} = 742.8,\ z = \frac{742.8 - 792.5}{90/\sqrt{22}} = -2.59$$

$z$ lies in CR, so reject $H_0$ in favour of $H_1$. The mean determination from Michelson's experiments was significantly different from the 'true' value of the speed of light.

**2 a** Let $\mu$ cm be the population mean length of one-year-old boys in Glasgow. Assume the population standard deviation is 2.53 cm.

$H_0$: $\mu = 75.80$ vs $H_1$: $\mu \neq 75.80$

$$\bar{x} = 76.06,\ z = \frac{76.06 - 75.80}{2.53/\sqrt{67}} = 0.84$$

$$p = 2[1 - \Phi(0.84)] = 0.4010$$

This value of $p$ is quite high, so do not reject $H_0$. The population mean length of one-year-old boys in Glasgow is not significantly different from 75.80 cm.

**b** Let $\mu$ cm be the population mean length of one-year-old girls in Glasgow. Assume the population standard deviation is 2.41 cm.

$H_0$: $\mu = 73.99$ vs $H_1$: $\mu \neq 73.99$

$$\bar{x} = 74.13,\ z = \frac{74.13 - 73.99}{2.41/\sqrt{55}} = 0.43$$

$$p = 2[1 - \Phi(0.43)] = 0.6672$$

This value of $p$ is quite high, so do not reject $H_0$. The population mean length of one-year-old girls in Glasgow is not significantly different from 73.99 cm.

**3 a** Let $\mu_D$ kg be the population mean change in weight for controls. Assume the population standard deviation is 2.75 kg.

$H_0$: $\mu_D = 0$ vs $H_1$: $\mu_D < 0$

$$\bar{d} = -0.526,\ z = \frac{-0.526 - 0}{2.75/\sqrt{151}} = -2.35$$

$$p = P(Z \leq -2.35 \mid H_0) = 1 - \Phi(2.35) = 0.0094$$

This value of $p$ is very low, so reject $H_0$ in favour of $H_1$. In the control population, there is a significant mean loss of weight.

**b** Let $\mu_D$ kg be the population mean change in weight for treated patients. Assume the population standard deviation is 2.75 kg.

$H_0$: $\mu_D = 0$ vs $H_1$: $\mu_D < 0$

$$\bar{d} = 0.431,\ z = \frac{0.431 - 0}{2.75/\sqrt{142}} = 1.87$$

$$p = P(Z \leq 1.87 \mid H_0) = \Phi(1.87) = 0.9693$$

This value of $p$ is very high, so do not reject $H_0$. In the control population, there is not a significant mean loss of weight.

**c** Prescription of a nutritional supplement may help to avoid weight loss during hospitalisation of elderly people.

## PREPARATION FOR UNIT ASSESSMENT

**(page 131)**

**1** 0.034

**2** **a** $\mu = 26, \sigma^2 = 81$ **b** $\mu = 2, \sigma^2 = 13$

**3** 0.0729

**4** 0.1048

**5** $31.08 \approx 31$

**6** Cluster sampling

**7** $\bar{x} = 105.4\,\text{mm},\ s^2 \approx 126.8\,\text{mm}^2$

**8** $H_0: \mu = 103$ vs $H_1: \mu < 103$
One-tail test
$p$-value $= 0.0122 < 0.05$
Reject $H_0$ at the 5% level.
There is evidence that adjustment has reduced the contents of jars.

## PREPARATION FOR COURSE ASSESSMENT

**(page 133)**

**1** 0.6

**2** $\frac{1}{3}$

**3** **a** $X \sim \text{Bin}(10, 0.25)$ **b** 0.0162

**4** 0.1357

**5** **a** 0.4170 **b** 15

**6** 0.0125

**7** **a** (41.04, 44.96)
**b** The CI contains 44 g so we would accept that the true mean could be 44 g.

**8** (0.24, 0.56)
99% of such intervals will capture the true population proportion, so it is very likely that between 24% and 56% of pupils took a foreign holiday in the past year.

**9** Stratified random sample:
list boys' names and number 1 to 250.
Use 25 distinct random numbers in the range 1 to 250 and select the boys with these numbers.
List girls' names and number 1 to 350.
Use 35 distinct random numbers in the range 1 to 350 and select the girls with these numbers.
Simple random sample:
number all pupils from 1 to 600.
Use 60 distinct random numbers in the range 1 to 600 and select the pupils with these numbers.
With simple random sampling, boys and girls in the sample may not be in the same proportion as in the population (school).

**10** Difference = brand A − brand B
$H_0: \mu_D = 0$
$H_1: \mu_D > 0$ One-tail test
$\sigma_D = \sqrt{35^2 + 35^2} \approx 49.5$
$z = \dfrac{\bar{d} - \mu_0}{\sigma_D/\sqrt{n}} = \dfrac{5 - 0}{49.5/\sqrt{8}} \approx 0.29$
$p$-value $= P(Z \geq 0.29) = 1 - \Phi(0.29) = 0.3859$
There is insufficient evidence to reject $H_0$.
The oil company's claim is not justified.

# Index

answers 146-63

bar diagrams 83, 105
BASIC 104
Bayes' Theorem 23-6, 30
bias, selection 79
Binomial distribution 42-5, 50, 63-4, 70-2
  tables 135-7

calculators 15, 45, 49, 55, 61, 69, 78, 126
census 75
Central Limit Theorem 82-4, 86, 104-5, 107
circles worksheet 78-9, 141
cluster sampling 96-7, 108
combinations 14-15, 29
complements (of events) 4, 29
computer program 104-5
conditional probability 18-20, 30, 117, 128
confidence interval (interval estimate) 100
  for population mean 86-7, 107, 125-6, 129
  for population proportion 90-2, 108
continuity correction 64, 72, 83, 84
continuous random variables 51-4, 61, 71
convenience sampling 98-9, 108
critical region 120, 121, 125, 129
critical value 119, 120, 129
cumulative distribution function (cdf) 33, 52, 53-4, 71
cumulative probabilities 43

data, paired 122-3, 129
data tables 142, 143-4
diagnostic testing 28
discrete random variables 33-4, 39, 51, 70
distribution of sample means 79-81, 106
  discrete 105
  uniform 83

empty sets 4, 29
estimating 114
estimator, unbiased 80, 145
events 3-5, 29
  independent 19-20, 30
  mutually exclusive (disjoint) 5, 9, 29
evidence, assessing 113, 115, 117-18, 119-21
expected value 70, 80
experiments 29
  Darwin's 122-3
  random 1-2, 3-4, 29

factorials 13
family size data 142
formulae 34, 42, 43, 44, 47

histograms 52
hypothesis
  null 112-15, 117-18, 119-21, 122-3, 124-6, 128
  study (alternative) 111-15, 117-18, 119-21, 122-3, 124-6, 128
  testing 111-30

independent random variables 39, 40-1
  and identically distributed (iid) 79-80, 82, 106-7, 145

Law of Total Probability 23-6, 30
laws of expectation and variance 37-41, 61, 70

modelling data 66-9
multi-stage cluster sampling 97

natural variation 79, 80
normal distribution 57-9, 72
normal random variables 72
null distribution 115, 128
null hypothesis 112-15, 117-18, 119-21, 122-3, 124-6, 128

one-stage cluster sampling 108
one-tailed alternative hypothesis 118, 120, 122, 128-9
outcome 2, 4, 5, 29
  equally likely 10-11, 13, 29

*p*-value 117, 118, 119, 123, 125, 128
paired data 122-3, 129
permutations 13-15, 29
Poisson distribution 46-7, 49, 50, 70
  tables 138
population 75, 76, 106
  mean 86-7, 106, 107, 114-15, 128
  parameter 75, 79, 80, 106, 112, 128
  proportion 90-2, 108, 127, 128
  variance 145
preparation for assessment 131-2, 133-4, 163

probability 1-32
conditional and unconditional 18-20, 25, 28, 30, 117, 128
curves 52-3, 57-8
density function (pdf) 52, 53, 54, 71, 72, 77
distribution 33, 37-40, 42, 70
plot 68-9
rules 7-9
proportional allocation 95

QBASIC 104
quizzes 66-7
quota sampling 99, 108

random experiments 1-2, 3-4, 29
random numbers 78
random sampling 77-8, 79, 106
stratified 94-5, 97, 103, 108
random variables 33-74
continuous 51-4, 61, 71
discrete 33-4, 39, 51, 70
independent 39, 40-1
and identically distributed (iid) 79-80, 82, 106-7, 145
normal 72
rules, probability 7-9, 19, 20, 63, 72

sample space 2, 4, 23, 29
sample statistic 106
samples 75, 76, 79, 106
biased 76, 79
sampling 75-110
cluster 96-7, 108
convenience 98-9, 108
distribution 80, 102
fraction 76, 106
frame 76, 78, 106
probability 77
reasons for 75
simple random 77-8, 79, 106
stratified random 94-5, 97, 103, 108
systematic 99, 108
units 76, 78, 106
variation 79

significance level 119, 120, 125, 129
speed of light 130
spreadsheets 78, 104
standard deviation 34, 80-1, 91, 107, 114
standard normal distribution ( ) 58, 72
table 139, 140
standard normal random variable 72
statistical inference 76, 77, 106
statistics in action 28, 66-9, 102-5, 127
stem and leaf diagrams 68
strata 97
stratification variables 94-5
stratified random sampling 94-5, 103, 108
study hypothesis (alternative hypothesis) 111-15, 117-18, 119-21, 122-3, 124-6, 128
systematic sampling 99, 108

tables 135-40
temperature conversion 56
test of population mean 114-15
test statistic 115, 117, 119, 128
tree diagrams 24-5
trials 2, 4, 29
two-stage cluster sampling 97
two-tailed hypothesis 117-18, 119, 120, 128-9

uniform distribution 83

variables, random *see* random variables
variance 34, 38, 39, 40, 43, 70, 145
variation, natural 79, 80
variation, sampling 79

wheat yield 102-3, 143-4

*z* test 119-21, 122, 123, 128

# Acknowledgements

The authors and publishers are grateful for permission to reproduce the following copyright material:

**Circles worksheet** (page 141)
used in Exercise 2 of Chapter 3

**Frivolous Pursuit Quiz** (page 66)
used in Chapter 2, Statistics in Action, True/False quizzes

Both are from: Mary Rouncefield and Peter Holmes, *Practical Statistics*, Macmillan 1989
© Crown copyright 1989. Published by permission of the Controller of Her Majesty's Stationery Office

The authors and publishers would also like to acknowledge the following sources of statistics used in the text:

**Growth and growth standards**
(page 116)
Savage, S.A.H., Reilly, J.J., Edwards. C.A., Durnin, J.V.G.A. (1999) Adequacy of standards for assessment of growth and nutritional status in infancy and early childhood, *Archives of Disease in Childhood*, **80**, 121–124

**Cross-fertilised and self-fertilised plants**
(page 122)
Darwin, C. (1876) *The Effect of Cross- and Self-fertilization in the Vegetable Kingdom*, 2nd edn, London: John Murray

**Lung function testing of children with cystic fibrosis**
(page 124)
Whiteford, M.L., Wilkinson, J.D., McColl, J.H., Conlon, F.M., Michie, J.R., Evans, T.J. and Paton, J.Y. (1995) Outcome of *Burkholderia (Pseudomonas) cepacia* colonisation in children with cystic fibrosis following a hospital outbreak, *Thorax*, **50**, 1194–1198

**Measurement of fat-free mass in children**
(page 124)
Reilly, J.J., Wilson, J., McColl, J.H., Carmichael, M., Durnin, J.V.G.A. (1996) Ability of biolectric impedance to predict fat-free mass in prepubertal children, *Pediatric Research*, **39**, 176–179

**WOSCOPS study**
(page 127)
Shepherd, J., Cobbe, S.M., Ford, I., Isles, C.G., Lorimer, A.R., Macfarlane, P.W., McKillop, J.H., Packard, C.J. (1995) Prevention of coronary heart-disease with Pravastatin in men with hypercholesterolemia, *New England Journal of Medicine*, **333**, 1301–1307

**Energy supplementation of elderly hospital patients**
(page 130)
Potter, J.M., Roberts, M.A., McColl, J.H., Reilly, J.J. (2001) Protein energy supplements in unwell elderly patients – a randomised controlled trial, *to appear*

**Wheat yields**
(page 102)
Wiebe, G.A. (1935) Variation and correlation among 1500 wheat nursery plots, *Journal of Agricultural Research*, **50**, 331–357

Every effort has been made to trace all the copyright holders, but where this has not been possible the publisher will be pleased to make any necessary arrangements at the first opportunity.

The authors acknowledge with gratitude the advice and encouragement of their families, colleagues at Kyle Academy Ayr, and at the University of Glasgow.